Symphony of Matter and Mind

Part Two

THEORY OF ENERGY HARMONY

Mechanism of Fundamental Interactions

Stanislav Tregub

TREGUB S.V.

Copyright © 2021 Stanislav V. Tregub

All rights reserved

No part of this book may be reproduced, or stored in a retrieval system, or transmitted in any form or by any means, electronic, mechanical, photocopying, recording, or otherwise, without express written permission of the author and publisher, except for short citations in relevant context.

For information about permission to reproduce selections from this book, please, write to symphony@stanislavtregub.com

Symphony of Matter and Mind. Part Two.
Theory of Energy Harmony. Mechanism of Fundamental Interactions.

by Stanislav Tregub

ISBN 9798451029459

Cover design by: Stanislav Tregub

The author is not responsible for the websites to which there are links in the book, and does not guarantee that the content of these sites will remain intact and relevant to the topic.

To my father

TABLE OF CONTENTS

Introduction ix
Chapter 1 The Fundamental Mystery 1
Chapter 2 Back to Basics 13
Chapter 3 The Hydrodynamic Electromagnetism 30
Chapter 4 The Puzzle of Gravitation 62
Chapter 5 Solving the Puzzle 92
Chapter 6 The Nuclear Muddle 113
Chapter 7 The Way out of the Muddle 139
Chapter 8 The Harmonization of Chaos 164
References 179

INTRODUCTION

Current mainstream theories of physics, the Standard Model of particle physics and the General Theory of Relativity, are incompatible due to the different mechanisms that they offer as explanations of fundamental energy interactions. This is considered the main problem for unifying them into a 'theory of everything.' Unfortunately, the problems are not limited to this issue.

Both theories contain arbitrary variables and constants that do not have any physical meaning and are fitted to the results of experimental tests every time the predictions fail. Moreover, the equations lead to infinities that are hidden by mathematical tricks to adjust the solutions to reality. Many physicists consider this internal inconsistency to be a sign of the mathematical ingenuity of the models. However, the sad truth is that the descriptive and explanatory basis of the models is a muddle and the predictive power is zero. Thus, they are practically useless. On top of this, both postulate the existence of virtual entities responsible for observable physical interactions. This means that the models have become metaphysical belief systems. Some physicists dare to correctly call the situation the fall of theoretical physics as a science. To see it rise, we need an alternative path.

In the second volume, the author continues to build the Theory of Energy Harmony based on the model of the universal mechanism proposed in the first part of the study. This mechanism underlies all fundamental interactions and can be called a unifying physical principle. The model does not use any virtual "ghosts" or arbitrary postulated parameters. It is self-consistent and adequate to reality. It contains only empirically verifiable assumptions and predictions. This is a paradigm shift that takes us back to physics.

CHAPTER 1

THE FUNDAMENTAL MYSTERY

To tell us that every species of things is endowed with an occult specific quality by which it acts and produces manifest effects, is to tell us nothing.

Isaac Newton

People noticed a long time ago that in nature there are two main and opposite directions of interaction: attraction and repulsion, merger and separation, creation and destruction. They attributed a religious character to this phenomenon and created mystical entities responsible for it. These were the gods of creation and destruction, or one god became two-faced and combined two opposite principles. Then came the era of the scientific approach, and physicists began to call the phenomena of interaction manifestations of "forces." They gave different names to these forces (electromagnetic, gravitational, strong nuclear, weak nuclear). However, in order for such an approach to differ from a religious one, an explanation of the physical mechanism of the work of such forces is needed. Otherwise, there is no difference between the concept of Force and God, since the paths of both are inscrutable. The concept becomes an auxiliary variable that patches a hole in the explanatory base and can act as it pleases.

There is a very subtle point here. If we use the verb "act," then its personal form carries the connotations of the work of some transcendent entity that acts on the phenomena of the world in one way or another. This again brings us back to the religious concept of the world order. It is no coincidence that physicists gradually switched to the term "interaction." This word denotes a process that occurs with the phenomena themselves and is carried out by them. However, to say that a fundamental interaction occurs is just as insufficient as to say that a fundamental force is at work. We need to understand the mechanism of interaction.

Mainstream theoretical physics applies the "good old" corpuscular model, which can be briefly described as follows: everything complex consists of simple parts; simple parts in their depths consist of the simplest indivisible parts. This is reflected in the name of the main theory of physics: the Standard Model (SM) of elementary particle physics. It is the heiress of the Quantum Mechanics (QM) model that we have dealt with in detail in the first part of the "Symphony of Matter and Mind" series.

The question remains: how do those fundamental parts interact to form complex things? Here is a short description of the SM answer: all interactions are the exchange of virtual particles. Such a description involves virtual entities that somehow take part in the real physical process and are the force behind it. The ghosts materialize the world: they are responsible for the interaction of physical entities and for the formation of these entities. Astonishingly, this loss of physical meaning is taken as a standard in theoretical physics.

It is very important to remember here that we are talking about interaction at a distance because no matter how much space there is between particles, it still exists. For example, if we correlate the scales of the Solar system and the atom, the electron in the nearest orbit in the hydrogen atom will be 12 times farther than Pluto from the Sun or 490 times more distant than the Earth from the Sun (Hotson, 2002). How does SM explain such interaction at a distance? The answer looks like this: virtual particles (photons, gluons, bosons, gravitons) are carriers of the corresponding interaction (electromagnetic, strong, weak, gravitational) and constitute the fields corresponding to these interactions.

For example, here is how any textbook on particle physics describes the mechanism of the strong interaction: quarks exchange gluons. The question from a student: what are quarks? The answer from SM: quarks are structureless point particles that no one has ever observed but we know that they exist as they constitute hadrons (protons and neutrons). The question from a student: how did we determine that they are there if they are unobservable? The answer from SM: they should exist because they explain the laws of the relationships in the interactions of other particles.

Physicist Fritjof Capra wrote: "The persistent failure to detect them experimentally, plus the serious theoretical objections to their existence, have made the reality of quarks extremely doubtful. On the other hand, the quark model continues to be very successful in accounting for the regularities found in the particle world ... The quark puzzle has all the traits of a new koan, which, in turn, could lead us to a major breakthrough in our understanding of subatomic particles" (Capra, 1975). We can only wonder, how a koan (a statement that has no logical structure, contains alogisms and paradoxes) can lead to a breakthrough.

It is interesting that within the particle zoo there is even such a beast called "Kaon" (K meson). It is a group of four mesons with a quantum number called "strangeness." They are also called bound states of a "strange quark" (or antiquark). They are used as an explanation of the matter-antimatter symmetry (Nobel Prize in 1980). They are at the foundation of the quark model within SM and the theory of "quark mixing" (Nobel Prize in 2008). This strange mix of virtual quark koans and kaon koans seems to be good foundation for real prizes.

Gluon is another koan of mainstream physics. It is postulated as an elementary particle without mass, which combines everything like glue (hence the name). It is a carrier of the strong interaction between quarks. According to SM, strong nuclear interaction, which "glues" nuclei together, behaves strangely. "Up to a distance of around a fermi (10^{-15} m.), it is very strongly repulsive, keeping the nucleons apart. Then, for no apparent good reason, it changes abruptly to very strongly attractive, then drops off very rapidly, so that at a distance of around three fermis it becomes immeasurable. This peculiar shape has never been successfully modeled by any theory. Note how current theory, in which the fudge is an accepted scientific procedure, "solves" this problem. Since current theory can't model this *observed* force, it simply ignores it, and instead invents (fudges) an *unobserved* (fifth!) force carried by eight "gluons" (*designed to be unobservable*) between eighteen or thirty-six "quarks" (*also designed to be unobservable*) inside the nucleon. It then "suggests" that this fudged gluon force in some unspecified way "leaks out" of the nucleon to make up the peculiar shape of the measured strong force ... After fifty or so years of effort, the huge physics establishment admittedly has failed utterly to provide a model that comes close to matching that peculiar shape of the nuclear force" (Hotson, 2002).

Here are more questions from a student who has not traded physical sense for good scores on the exam. How do these virtual particles without mass and structure convey interaction? Answer: they fly in the void on their angelic (massless) wings and do what they should. Question: let the angels bring the "glad tidings," but how do they transmit it? Answer: a real elementary particle swallows the virtual angel. How does it swallow? Isn't it elementary, structureless, fundamental, indivisible, and so on? And where did these angels without flesh come from? Answer: another real particle spat them out. How? Isn't it elementary, structureless, fundamental, indivisible, and so on? Sit down, student. You poorly studied the introductory dogma: everything happens through the interaction of particles. If there are no particles, we will invent them and their special properties to keep a lid on things. Particles copulate, give birth, or swallow each other, and let it remain their intimate secret. The main thing is to describe the process with beautiful formulas and diagrams. Shut up and calculate.

The student sits down and thinks: you need to either leave the mainstream and look for physical meaning or believe in ghosts, but get the best marks, and then a salary, titles, bonuses and prizes. Generations of students made the second choice. As a result, the Standard Model is the mainstream in theoretical physics. But in essence it is a belief system that is similar with the most ancient religious picture of the world — animism. According to this model, there are many virtual spirits in the world, which are responsible for the phenomena observed in it. This version of the mechanism has existed for thousands of years. But it is not about a physical mechanism and is not about science.

However, the field of knowledge we are talking about is called elementary particle physics and it claims to be scientific. Any textbook or encyclopedia will necessarily contain lines stating that the SM is the most successful and experimentally proven theory in the history of science. What is its success? The mathematical description of virtual objects. However, the question of the physical reality of the interactions remains.

Capra wrote: "Particles are processes rather than objects … The simple mechanistic picture of basic building blocks had to be abandoned, and yet many physicists are still reluctant to do so. The age-old tradition of explaining complex structures by breaking them down into simpler constituents is so deeply ingrained in Western thought that the search for these basic components is still going on … The incomplete character of a theory is usually reflected in its arbitrary parameters or 'fundamental constants,' that is, in quantities whose numerical values are not explained by the theory, but have to be inserted into it after they have been determined empirically. Quantum theory cannot explain the value used for the mass of the electron, nor field theory the magnitude of the electron's charge, or relativity theory that of the speed of light. In the classical view, these quantities were regarded as fundamental constants of nature which did not require any further explanation. In the modern view, their role of 'fundamental constants' is seen as temporary and reflecting the limitations of the present theories. According to the bootstrap philosophy, they should be explained, one by one, in future theories as the accuracy and scope of these theories increase. Thus, the ideal situation should be approached, but may never be reached, where the theory does not contain any unexplained 'fundamental' constants, and where all its 'laws' follow from the requirement of overall self-consistency" (Capra, 1975).

Capra is right: nothing in physics, as, indeed, in any branch of science, is ideal. Any theory can have only the best explanatory power at this stage of cognition and cannot claim to be final. The new idea may just be a hypothesis, or it may be a whole set of hypotheses and a coherent, self-consistent theory. It can never be entirely new. Any theory is a bootstrapping from the previous ones, even if it refutes some of them. Nevertheless, it must possess the primary feature of a scientific model: to be testable and refutable in principle. It cannot contain arbitrary parameters or virtual entities that "explain" things while being unexplainable themselves.

Unfortunately, the situation in theoretical physics has not changed, and everyone is still talking about some future theory that should explain all of the conundrums self-consistently. For example, physicist and psychologist Arnold Mindell wrote: "If you don't like the concept of virtual particles, you can always come up with a better idea. You never have to understand or agree with anything in physics. If you can think of something better, just rebel against what has been taught until now, and prove your ideas. Nothing in physics is final. If something in you refuses to understand something in physics, it may be that everybody else is wrong and something is missing" (Mindell, 2000).

Mindell is correct: if you do not agree to something in current theories, just offer your ideas and prove them. It makes no sense to just criticize and negate without suggesting anything instead. But as a psychologist, he must understand that rebellion is not as easy as he describes. In reality, it is suicide for a physicist's formal career. It is impossible to make a career if you don't wear the "Particle" team uniform. You should not think, but just play for your team. The "Particle" team sets the rules. There are only two of them. Rule number one: the "Particle" team is always right. Rule number two: if it is not right, see rule number one.

The history of science teaches us two lessons. First, old paradigms are not easily discarded. There are multiple factors that contribute to this phenomenon, such as politics, economics, personal beliefs, and national interests. However, the main reason is that the model of reality must be both stable and dynamic. If the model is too dynamic, it cannot meet the current tactical needs. Conversely, if the model is too rigid, it cannot provide the strategic needs of adaptation to a dynamic world, and eventually, it will simply collapse. Second, history teaches us that a paradigm shift is inevitable but only after a decisive accumulation of evidence in favor of a more accurate representation of the world. Until then, no matter how many contradictions the old model contains, it will remain the dominant theory.

Physicist Matt Strassler described the problems of the corpuscular model as follows: "The term "virtual particle" is an endlessly confusing and confused subject for the layperson, and even for the non-expert scientist. I have read many books ... and all of them talk about virtual particles and not one of them has ever made any sense to me ... The best way to approach this concept, I believe, is to forget you ever saw the word "particle" in the term. A virtual particle is not a particle at all. It refers precisely to a disturbance in a field that is not a particle. A particle is a nice, regular ripple in a field, one that can travel smoothly and effortlessly through space, like a clear tone of a bell moving through the air. A "virtual particle," generally, is a disturbance in a field that will never be found on its own, but instead is something that is caused by the presence of other particles, often of other fields. Analogy time (and a very close one mathematically); think about a child's swing. If you give it a shove and let it go, it will swing back and forth with a time period that is always the same, no matter how hard was the initial shove you gave it. This is the natural motion of the swing. Now compare that regular, smooth, constant back-and-forth motion to what would happen if you started giving the swing a shove many times during each of its back and forth swings. Well, the swing would start jiggling around all over the place, in a very unnatural motion, and it would not swing smoothly at all. The poor child on the swing would be furious at you, as you'd be making his or her ride very uncomfortable. This unpleasant jiggling motion — this disturbance of the swing — is different from the swing's natural and preferred back-and-forth regular motion just as a "virtual particle" disturbance is different from a real particle. If something makes a real particle, that particle can go off on its own across space. If something makes a disturbance, that disturbance will die away, or break apart, once its cause is gone. So it's not like a particle at all, and I wish we didn't call it that" (Strassler, n.d.)

So, particles are not particles at all but ripples in the field. Some ripples are nice and regular so they are real, others are disturbed ripples and they are virtual. But here come the main questions of theoretical physics. How are those nice or disturbed structures produced? How do they interact? The authors explanation: something makes them and they go off across space; if this something as the cause of everything is gone, they die away. This explanation makes no sense to either the layman or the non-expert scientist, unless that person is fascinated by magical explanations of how something produces something with the shove of a hand

(God's?), that makes those nice regular motions of the swing or causes them to jiggle.

Does it make sense to the expert scientist such as Strassler? It seems so because he does not care to dwell on the mechanism of the creation any further and proceeds with the explanation of the interaction mechanism: "Two electrons approach each other; they generate a disturbance in the electromagnetic field (the photon field); this disturbance pushes them apart, and their paths are bent outward. One says they "exchange virtual photons," but this is just jargon ... But if two electrons pass near each other, they will, because of their electric charge, disturb the electromagnetic field, sometimes called the photon field because its ripples are photons. That disturbance is not a photon. It isn't a ripple moving at the speed of light; in general isn't a ripple at all ... This disturbance is important, because the force that the two electrons exert on each other — the repulsive electric force between the two particles of the same electric charge — is generated by this disturbance. The same is true if an electron and a positron pass near each other; the disturbance in this case is similar in type but different in its details, with the result that the oppositely charged electron and positron are attracted to each other" (Ibid).

Things interact by the magic of the disturbance of the field which is a ripple called photon which is not a ripple at all. This is an enlightening explanation. To top it off with a nice cherry, it all happens because of the charge type. How? Just because there is a charge sign that makes things attract or repel depending upon whether it is opposite or similar. Why do these charge types attract or repel? Because they are of a specific type. The author does not even notice that this is a circular logic (logical fallacy) and considers that this completes the explanation of the mechanism: "That said, it is not at all mysterious; it is something whose details, if we know the initial motions of the electrons, can be calculated easily. Exactly the same equations that tell us about photons also tell us about how these disturbances work ... Perhaps unfortunately, this type of disturbance, whose details can vary widely, was given the name "virtual particle" for historical reasons, which makes it sound both more mysterious, and more particle-like, than is necessary (students of math and physics will recognize real photons as solutions of a wave equation)" (Ibid).

Students of mathematical physics, who must shut up, calculate and not care about physical sense, remember that the wave equation was stripped of its physical meaning as the description of waves in a medium and turned into the description of the probability of finding particles in this or that state (Copenhagen interpretation of quantum mechanics). Overall, the quantum equations do not tell us how these disturbances work. The attempts to describe the mechanism of interaction come to the same old mantra of exchange of virtual particles that the author calls just jargon. They substitute the phenomena of energy oscillations in a medium with magic interactions of particle-like entities in an empty space.

The author tries to explain: "Physicists often say, and laypersons' books repeat, that the two electrons exchange virtual photons. But those are just words, and they lead to many confusions if you start imagining this word "exchange" as meaning that the electrons are tossing photons back and forth as two children might toss a ball. It's not hard to imagine that throwing balls back and forth might generate a repulsion, but how could it generate an attractive force? The problem here is that the intuition that arises from the word "exchange" simply has too many flaws. To really understand this you need a small amount of math, but zero math is unfortunately not enough. It is better, I think, for the layperson to understand that the electromagnetic field is disturbed in some way, ignore the term "virtual photons" which actually is more confusing than enlightening, and trust that a calculation has to be done to figure out how the disturbance produced by the two electrons leads to their being repelled from one another, while the disturbance between an electron and a positron is different enough to cause attraction" (Ibid)

The mechanism of "virtual particles exchange" is just words that cause confusion. But is there any alternative explanation? No, we do not need one because what we need is just math to figure out how things attract or repel. The standard way of the Standard Model is to confuse a physical explanation with a mathematical description, trust in it and get enlightened.

This is how the enlightenment sounds: "The language physicists use in describing this is the following: "The electron can turn into a virtual photon and a virtual electron, which then turn back into a real electron." And they draw a Feynman diagram (which is) a calculational tool, not a picture of the physical phenomenon ... For those who learned (and recall a bit of) freshman physics, what is happening is that the oscillating electric field that makes up the photon is polarizing the electron field — inducing a dipole moment. Remember dielectrics and how electric fields can polarize them? Well, the vacuum of empty space itself, because it has an electron field in it, is a polarizable medium — a dielectric of sorts" (Ibid).

We need to stop here because the picture gets so confused that Strassler does not even notice the oxymorons that it produces. Just look at the last statement: the empty space is a polarizable medium. How can the void be a medium? If it is a medium, it is not an empty space. This is what any person who recalls a bit of freshman physics and has not parted with common sense would think. Are there any reasons the expert's explanation gets so confused? There are two major reasons.

The first one is political: the mantra about the empty space is still so imprinted in the minds of mainstream physicists (experts) that they do not even notice how they reach a dead-end using it and begin to contradict themselves. It started with Einstein abolishing the notion of an all-encompassing medium: "The first step we must take is to *give up the ether* ... From this it follows that the only way to arrive at a satisfactory theory is to give up the notion of a medium filling all the space. This is the first step to be taken" (Einstein, 1910). What did he leave physicists with? "The region of space without matter and without an electric field appears completely empty, i.e., it cannot be characterized by any physical quantities" (Einstein, 1918).

Something that cannot be characterized by physical parameters is an intangible void. Did he or others arrive at a satisfactory theory using the concept of a void?

Not at all. That is why there is still no unified theory of fundamental interactions. By the end of his life, Einstein had to admit that "the concept empty space loses its meaning" (Einstein, 1954).

We will deal with these vicissitudes of his model in the later chapters where we will make a step back and give up the void to return physical meaning to our modeling of the world. Here we only note that the intuition about wave phenomena, which Strassler refers to, does not work for phenomena occurring in the void, because waves, by definition, are the movement of oscillations in a medium. That's why he has problems when he tries to unite the empty space and the field. If there is an empty space in-between, how can something induce momentum on something? We can only think of some virtual entities flying in it and carrying the interaction momentum on their "angelic wings." The indirect interaction of real things at a distance becomes possible only by exchanging some unreal things as only something intangible can move in an intangible void. But such a picture requires faith, not scientific knowledge.

So, the second reason for the confusion is related to the main question of theoretical physics: how matter forms and interacts. If we think of the mechanism as some magic transformation of virtual entities into real entities, we are back to the religious animism that speaks of spirits. The only difference is that SM offers a mathematical description of these spirits. The author speaks the truth: a diagram of particle interaction is just a math tool not a picture of a physical phenomenon.

But we are back to the question of attraction and repulsion: "It's not hard to imagine that throwing balls back and forth might generate a repulsion, but how could it generate an attractive force?" (Strassler, n.d.) The author tries to explain it by ripples: "The reason ties back to the very reason that there are anti-particles in the first place: every field, by its very nature, has particle ripples and anti-particle ripples" (Ibid) This is the explanation: the field by its nature has particles and anti-particles that attract or repel. Why? Because they are particles and anti-particles. The circle is closed. We are enlightened.

The author wishes that we would not call those ripples "particles" but keeps on using the term for historical reasons. Of course, the word inevitably leads to an image of a discrete entity, not a ripple (continuous phenomenon). This intuition is the basis of the millennia-old atomistic paradigm. It certainly has many flaws. The main one is that it gets into an impasse when it comes to the question of the mechanism of interaction. It inevitably leads to a mechanical picture of the process as flying balls that repel or attract. This picture can describe direct contact. But when it comes to interaction at a distance it is useless. Moreover, it cannot explain both sides of the interaction. And we should not forget the balanced state due to which stable structures exist. The only way to arrive at a satisfactory theory is to account for all three aspects (attraction, repulsion, balance) and return to the notion of a medium filling all the space.

Often people say that the problem is not in terminology. Yes, we can choose any words. But words carry meanings. We lose meaning if we are careless with the terms. Can we choose another word to describe the fundamental levels of matter? Can we use another analogy that will lead us to true enlightenment out of the oxymorons of SM and get us back to the physical sense? Werner Heisenberg, one of the pioneers of QM and its heiress SM, explained: "The only thing we know from the start is the fact that our common concepts cannot be applied to the structure of the atoms" (Capra, 1975). The creator of virtual particle interaction diagrams, Richard Feynman wrote: "I think I can safely say that nobody understands quantum mechanics ... Do not keep saying to yourself, if you can possibly avoid it, 'But how can it be like that?' because you will get 'down the drain', into a blind alley from which nobody has yet escaped. Nobody knows how can it be like that" (Feynman, 1965).

So, the prophets know from the start: do not even dream of any physical analogies; forget about common sense and physical meaning. Asking questions will not lead you to enlightenment but only into a blind alley. QM became a select branch of "super-knowledge" that no one understands, including the model's creators. Believing is the only solution for the parishioners of this church. But to get enlightenment we need to do exactly the opposite: start asking questions and seeking answers that have physical sense.

Let's go back to the musical analogy used by Strassler: the tone of a bell. Unfortunately, he does not pay enough attention to it. However, it has two main points that can lead us out of the difficulties of theoretical physics as a whole and the author in particular. First, it is a wave phenomenon, which can be described using common physical concepts. Second, it clearly indicates the need for a medium for the distribution of energy and the implementation of energy interactions. The author writes about the field all the time, but it turns out to be "empty space" for him. Thus, he nullifies his own analogy with the propagation of the sound of a bell in the air. Nothing can propagate in the void, neither "nice ripples" nor "disturbances."

As soon as we include an all-filling energy environment in the model, there is no need to build castles in the void. Moreover, the concept of virtual intermediaries of energy exchange becomes superfluous, no matter how we call them, "particles" or "disturbances." The all-filling environment is a universal intermediary. In fact, the word "intermediary" will not be correct because it is not just in between the material entities. It is what constitutes the entities and what is between them. We can call it with various words (ether, field, etc.) but we should stick to the meaning of a continuous energy environment with no void gaps. There are just various states of this energy in different regions. A vacuum is not an "empty space." It is just a region where forms of matter are not created. The only thing we need to solve the basic problem of theoretical physics is to show the mechanism that creates them.

If we choose a musical analogy, we are physically grounded. Such a change of analogy seems trivial but leads to a way out of the blind alleys of mainstream models of theoretical physics. We can explain things (even the structure of an atom) by common physical concepts. We can deal with all those mysteries that SM has proclaimed unexplainable. We return physical meaning to the wave equation that becomes the description of actual physical waves and not "waves of probability." We can explain the periodicity of the chemical elements table without using virtual balls orbiting other balls in the void. We can explain also the real "balls" of celestial bodies' movements and structures. We can explain the uncertainty principle without any mystical awe of Nature that plays tricks with us to make us uncertain about the position and momentum of balls in the void. We can explain the double-slit experiment result that Feynman called "the mysterious behavior in its most strange form" (Feynman et al., 1964-1966). But most important is that we can explain how structures of matter form and how they interact. It may sound like an unsupported extravagant claim. How can a simple change of an analogy perform such a "miracle"?

First, if the statements sound extravagant there is only one reason for that: usually, people take something new as extravagant. Only when they get acquainted with the idea it becomes self-evident and trivial. So, the reader should be patient, look into the details supporting the model and not jump to conclusions before that.

Second, we have dealt with the foundations of the model called the Theory of Energy Harmony (TEH) in the previous part of the study and solved the mysteries listed above using the musical analogy (see "Part One. Music of Matter"). So, the claims are supported and for an informed reader, they are, probably, not extravagant by now.

Third, we need to stress that the use of musical terms in TEH is not a loose metaphor but a strict physical analogy. Any scientific model is analogical as we build our new knowledge using the acquired knowledge of phenomena that demonstrate similar characteristics.

Fourth, simply using the concept of a clear tone (or better still, a musical note) as a general analogy does not solve the problem. It is necessary to elucidate the mechanism underlying the analogous system (in this case, musical phenomena) and the target system (in this case, fundamental energy processes). Moreover, saying that it is all about ripples (nice or disturbed) does not get us anywhere specific. We need to show how those ripples (or better still, energy oscillations) interact and form structures that we call atoms, molecules, chemical compounds, and further forms of inanimate and animate matter.

It is not enough to engage in hand-waving by describing the hand pushing of the swing to produce an even or jiggling motion. Even this mechanical analogy is not as simple as it seems. Let's get back to the description by Strassler: "Now compare that regular, smooth, constant back-and-forth motion to what would happen if you started giving the swing a shove many times during each of its back and forth swings. Well, the swing would start jiggling around all over the place, in a very unnatural motion, and it would not swing smoothly at all. The poor child on the swing would be furious at you, as you'd be making his or her ride very uncomfortable" (Strassler, n.d.).

There is a major gap in the description that prevents the theoretical physicist from explaining the real phenomenon. Thus, the analogy only raises further questions instead of giving the answer. Why should one shove make the swing do a "natural motion" and many shoves should cause "unnatural jiggling"? Actually, this is not true. The expert-theorist missed the mechanism of interaction of oscillations, and therefore its description does not correspond to reality. Any layman or even a non-expert scientist, who happened to help his child on a swing at least once in a life-time, knows that one shove is not enough for the child to be happy. The reason is again known to any layman: the swing loses energy due to friction. It will not go on swinging naturally and needs an input of energy. It could be an external shove or the child's own movement.

Here comes the main question: what shoves will make it continue its natural motion and which ones will cause it to jiggle? Any practical layman knows that for the ride to be comfortable one must synch with the swing motion parameters. It means that shoves should have a specific frequency related to the frequency of the swing and come at specific phases of the swing. If these two requirements are not met the shoves will cause a disturbance in the natural motion or even a complete stop. We are talking about frequency-phase coupling which is the essence of synchronization. We are talking about an optimal energy transfer mode and an interaction of oscillators. The swing and the hands make oscillations of different types and have different parameters. But for them to form a stable system they have to synch. They do not have to be in unison (1:1 frequency ratio). They do not have to be coupled constantly. They just need to intersect at some points in their phase space. They have to form a common phase portrait as a stable attractor. If they fall out of the synch region (specific ratios of their parameters), the system will collapse. This analogy works for any kind of interaction and energy exchange.

Back to the description of the expert: "This unpleasant jiggling motion — this disturbance of the swing — is different from the swing's natural and preferred back-and-forth regular motion just as a "virtual particle" disturbance is different from a real particle" (Ibid). Please, note that the author tries to explain the major tenant of SM that interactions are the result of virtual particles exchange. Ok, this is just jargon and we should call them disturbed ripples now. But, judging by the description of the author, disturbances are counterproductive for the movement of the swing. How can something counterproductive be the mediator of interaction? We are losing logic and common sense here. This is typical for the standard descriptions within Standard Model. Why? The reason is simple: they do not enlighten us on the actual mechanism of interaction.

Of course, going from the notion of a "virtual particle" to a "disturbed ripple of the field" makes the description sound more scientific as it gets rid of connotations of intangibility. But it does not help us in understanding the physical mechanism. This is why the author's formulation of a comparison of an analogous system (swing) with the target system (fundamental energy interactions) is incomplete, to put it mildly.

Let's get rid of the "virtual particle" or "disturbed ripple" concept in the above analogy. Let's imagine that the swing and a parent helping a child are interacting oscillations of energy. It is not a great stretch of the imagination as physically all forms of matter are structures of energy oscillations. For the interaction to be productive and create stable system these oscillations' parameters have to be within the synch region (frequency-phase coupling mode). If not, the swing will jiggle or even stop and the child will be furious.

Helping a child on a swing is a simple example of synch. It is far better to use the musical analogy because music is about the interaction of many oscillators with various parameters creating stable harmonious structures with specific patterns that can be described both by complex math of wave physics and simple math of frequency ratios that produce these structures. It really works as a physical analogy as it uncovers the universal mechanism of structure formation and energy interactions. As we have seen already in the previous volume of the series and will see later on, the description of this mechanism is quite intuitive and does not require any faith in virtual entities. To understand it we need a small amount of math to use it not only as a calculational tool but as a description of the physical phenomenon.

Here it makes sense to get back to the basic TEH hypothesis. It speaks of a mechanism that can be a realization of a dream to connect all Nature with the properties of whole numbers thus providing us with a model of the world that is coherent and simple. Simplex sigillum veri.

Hypothesis:

The Universe is a continuum of energy oscillations with various amplitude-frequency characteristics and phase portraits. These oscillations create forms of matter by frequency-phase coupling process (synchronization) which is optimal when parameters tend to harmonious ratios of integer numbers (synchronization region).

Using musical analogy, we can call fundamental stable structures of matter the musical notes. Composite forms are harmonious combinations of different notes that combine into the symphony of matter. In the previous part of the study, we developed detailed models of how notes of matter produce structures of the microworld (atoms, molecules, chemical compounds) and structures of the observable macroworld. Here we will move on to the mystery of energy interactions and the observed phenomena of attraction, repulsion and balanced state. For this, we need to take a step back and get to some freshman physics and look at it from a fresh point of view.

Chapter 2

Back to Basics

The key to all sciences is unquestionably the question mark. To the word How? we owe most of our greatest discoveries.

Honore de Balzac

Let's look at such a long-known and described by physics phenomenon as electricity. The electromagnetic interaction is called fundamental in physics for a reason: it is not only present throughout the known Universe but is the basis of our entire modern civilization. Without electricity and all its manifestations, we can hardly imagine our life. The practice of using this phenomenon is at a dizzying height. But the description and understanding remain at the level of calling for spirits.

We have to get back to the history of the issue. It is instructive, as there were crossroads where a dead-end was chosen. At the dawn of the study of electricity, it was called the intangible fluid (Franklin 1747). It sounded both physical (talking about the flow of something) and not physical (equating the current to other metaphysical entities, creating the spirit of electricity). In 1820, Hans Oersted showed that an electric current flowing through a circuit causes a deflection of the magnetic needle. The logical conclusion: if an electric current generates magnetism, then the appearance of an electric current may be associated with magnetism. The idea of turning magnetism into electricity captured Michael Faraday. For many years he did various experiments, and in 1831 a triumph came: he discovered the phenomenon of electromagnetic induction.

Faraday made a ring of soft iron and wound many turns of copper wire on each half of the ring. The wire closed the circuit of one winding, and there was a magnetic arrow in it. A current from a battery of galvanic cells passed through a second winding. When the current was turned on, the magnetic needle made several oscillations and calmed down. When the current was interrupted, the arrow oscillated again. To describe the observed magnetic lines and oscillation patterns of the arrow, he used the concept of a field and represented it as a set of lines of force emanating from sources (magnets, electrical wires). He went the usual way of explaining: if something moves, then there is a force acting; if something forms patterns of matter, it is a field at work.

If you put a magnet under a sheet of paper with iron fillings poured on it, lines form. There is a feeling of the presence of the field, which has these lines. You can also fix the magnetic compass needle's position at different points and make a description of the field, i.e., arrow positions connected by a line will be field lines. Similarly, you can measure the temperature difference in the room (the thermodynamic medium is not homogeneous) and describe this field in the form of lines. You can pour salt on a plate connected to an oscillation generator and see structured lines-patterns that depend on the oscillation frequency:

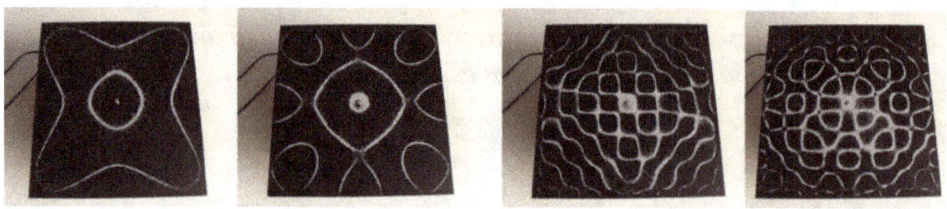

You can mix different chemical components as in the Belousov-Zhabotinsky reaction and get a visual picture of the vibrational processes in the Petri dish. You can place a colony of Dictyostelium discoideum amoebas in the same dish and observe the biochemical rhythm of intercellular communication:

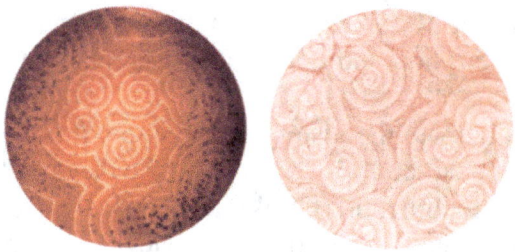

Left: Belousov-Zhabotinsky reaction. Right: a colony of bacteria (Epstein, 2006).

The obvious question arises: are these patterns the result of a process of interaction, or do they belong to some field? Do lines generate interaction or vice versa? Generations of physicists and many other people answered that fields create interaction. It turns out that the lines are not the result of interaction, but the interaction is the result of the presence of lines.

A follower of Faraday, James Maxwell, in 1864, derived a system of equations describing electromagnetic phenomena experimentally studied by Faraday and other scientists. Thus, the theory of electromagnetic fields was born. Here are examples of explanations in this theory. Question: why current flows through wires? Answer: because the field moves charged particles. Question: why does the compass needle move? Answer: because the magnetic field moves it. Question: why does the generator produce current? Answer: a change in the magnetic field inside the winding causes the appearance of an electric field that causes current. Why? Because it is the field. Why? Because.

But since the field was only a word for describing the phenomena of interaction, without explaining how this process occurs, it was necessary either to produce other entities or to repeat as a mantra: field-field-field. It happened both ways. At first, physics was "saved" by the measurability of the interaction results: it created the illusion of certainty. But then there were some perplexities. As the electric field and the magnetic field are interconnected (magnetic effects arose during the movement of electrical charges and, conversely, the movement of a magnetic charge created electricity), they were combined into one field. Two entities merged.

But there was a problem: if there was a movement of charges, then the field was created, and if not, then it was not present. The spirit was here, there and everywhere, and suddenly nowhere. When Maxwell faced this dilemma, he was ready to throw all his formulas describing the electromagnetic field into the trash bin. But then he calmed down and applied the standard method of "gluing" the theory together: for completeness of the formulas, he introduced an additional element (auxiliary variable), which in no way follows from the experiments. He created an additional member of the equation called the "bias current." The introduction of this quantity made it possible to close the system of equations of electrodynamics.

It was a classic trick of a mathematician who was supposed to reduce the equation so that it was complete and gave the impression of a model describing reality. A newly created entity, not being a physical phenomenon, not being a current, entered the equation and even acquired a unit of measure like a real current. Maxwell introduced the concept of the total current, where he added both the conduction current and the bias current. Discussions are still ongoing about where this "bias current" came from and why this variable was needed. Maxwell had one task: it was necessary to describe the interactions in the medium (he did not doubt its presence), but since this medium remained unknown, the interaction formula should include a variable reflecting some fundamental property of this medium (traditionally he called it ether). The name "bias current" is not accidental: it reflected a certain bias mechanism in an elastic medium that affects the electromagnetic interaction parameters and its propagation.

The mathematical trick created the illusion of harmony, but the same question arose: what is the mechanism of interaction? The concept of a field didn't help. Here is a standard description of our days: "An electromagnetic field … is a classical (i.e., non-quantum) field produced by moving electric charges. It is the field described by classical electrodynamics and is the classical counterpart to the quantized electromagnetic field tensor in quantum electrodynamics. The electromagnetic field propagates at the speed of light … and interacts with charges and currents … The field can be viewed as the combination of an electric field and magnetic field. The electric field is produced by stationary charges, and the magnetic field by moving charges (currents); these two are often described as the sources of the field" (Wikipedia, "Electromagnetic field").

Instead of clarification, confusion arises. Is the field the source of charges, or do the charges create the field? Does the field propagate and interact with charges, or charge currents propagate through the field? Such logical circles inevitably lead to a dead-end. This impasse of classical electrodynamics has led to the creation of a quantized field full of virtual particles that like angels transfer interactions on their wings. They are generated by one type of real particles and absorbed by another. Such a model is considered a way out of the impasse, but it is not. It only leads to the quantum analog of the classical dead-end: the need to invent new magical entities to explain various manifestations of interaction.

We can see lines of sawdust near the magnet and say: this is how the field works. We can put sawdust on a sheet that is under the influence of air vibrations (sounds), see the lines and say: here is the field. We can record the temperature readings of an inhomogeneous medium and represent it in the form of lines. It only remains for us to label these fields by different names: the spirit of temperature, the spirit of sounds, the spirit of electromagnetism, and so on. But concerning temperature or sound, we do not do it anymore. Why? The answer is simple: we understand the physics of the process in an air or other medium, which creates the observed effects of wave structures and their interactions. We know at a sufficient level of detail that oscillations in an environment give rise to the formation of structures, the transfer of energy, its concentration, the destruction of structures, energy dissipation, and so on through a chain of interactions that are no longer mystical to us. How big a leap of not even faith, but a probabilistic assumption is needed to extrapolate a similar mechanism to those aspects of reality where the medium is still not quite accessible for us — to the depths of the microcosm and the breadth of the macrocosm?

Let's try to make this leap. To do this, we have to abandon the oxymoron of the leading theories of modern physics: there is the void and there is something in it. We need to return to the primary hypothesis with a very long history: the whole Universe is an energetically filled, anisotropic and dynamic medium. Is this such a fantastic hypothesis? If we observe the energy environment wherever it is available to us, why can't we assume a medium at the measurement levels inaccessible to us? Yes, this medium may have different parameters; that is why we call it anisotropic and dynamic. This hypothesis means that we abandon the magical thinking of the childhood of humankind, forget about the mystical powers of spirits (regardless of the name), and recognize that physical causal relationships work in the physical world, though not all of its details are known to us. Such an idea would sound trite if leading theories of physicists remembered it, would not put inconsistencies in concepts under the rug, and would not fill them with hocus-pocus involving various spirits.

A simple question: what is electricity? Franklin spoke of the intangible fluid flowing in matter. Faraday said about the current in the field. If you ask a modern middle school student or a graduate of any university physics department, both will answer roughly like Wikipedia: "Electric charge is carried by charged particles, so an electric current is a flow of charged particles" (Wikipedia, "Electric current"). There is another option: "Electricity is the set of physical phenomena associated with presence and motion of matter that has a property of electric charge" (Wikipedia, "Electricity").

So, what is it? A movement of particles or movement of charge? At first glance, these are equivalent explanations, but there is one subtlety. They are identical only in the corpuscular paradigm. But the movement of a charge is not necessarily the movement of particles. In the wave interpretation, electricity becomes the distribution of energy, not the motion of objects. The charge is an energy concept. The second definition can be rephrased as follows: electricity is a set of physical phenomena associated with energy presence and motion. It will be so general that it will be about everything and about nothing at the same time. This is not enough to understand the phenomenon.

When interpreted in the Standard Model framework, the mechanism becomes the mysterious interaction of particles with their emission and absorption of one another with unknown details of the process, which are waiting for disclosure, but for now, remain the desired hidden variables. In the wave interpretation, the mechanism becomes the physical frequency-phase interaction of energy fluctuations with known variables. Let's try to consider both approaches and compare the result.

In SM, electricity is considered to be the movement of particles with a charge. The question arises: what is an electric charge? We read the same Wikipedia: "Electric charge is the physical property of matter that causes it to experience a force when placed in an electromagnetic field. There are two types of electric charge: *positive* and *negative* (commonly carried by protons and electrons respectively). Like charges repel each other and unlike charges attract each other" (Wikipedia, "Electric charge"). The charge is a source of fields, and fields, as a specific force, are a source of charge. Is it a cyclic causality or a logic cycle (logical error)? Rather, the second. What had to be explained becomes an explanation by itself.

What is an electric field? "A vector field surrounding an electric charge that applies force to other charges, attracting or repelling them ... (It) is created by electric charges or magnetic fields varying in time" (Wikipedia, "Electric field").

Does the charge create a field or the field generates a charge? Do you feel the ground slipping out from under your feet due to the dizziness from walking along a logical cycle? No? Let's try to cycle again: electricity is the movement of electric charges — it is an electric field — it is electric interaction — it is an electric force — it is electric current — it is electricity ... Should we continue? Many physicists seem to have a very stable vestibular apparatus. They spin in these cycles all school, student, doctoral, professoral and academic years.

Let the field be a collection of charges of interacting particles. But then the same question arises: how do they interact? To say that the charges are attracted and repelled due to a force of attraction and repulsion is the same as to the question "why does the car move?" answer "because the wheels spin." But the next question arises: why do wheels spin? If we are ignorant about the mechanism, it remains only to spin the "wheel" of logic and say that wheels are driven by the car, or create the illusion of stopping an endless circle and answer: such is their force. But the question is in the mechanism. The car moves not because it has a spirit of movement inside it but because it has a mechanism that creates this movement. When we apply such an analogy, the problem of cyclic description of electricity becomes obvious. But the fact is that theoretical physics has been spinning in the "wheel" of such a cyclic description for two centuries now.

Does this confuse mainstream propagandists? No. A physicist and popularizer of science, John Gribbin wrote in the Encyclopedia of Elementary Particle Physics: "Just as a car driver doesn't need to understand what goes on beneath the bonnet of the car in order to get from A to B, as long as quantum mechanics worked, you didn't have to understand it" (Gribbin, 2000). But even a child knows that if the mechanism under the bonnet breaks, the car will not reach point B. The driver may not know the mechanism, but he will not go anywhere if no mechanic understands the mechanism.

The fact is that the twentieth century's theoretical physics behaved like a careless driver who does not know what is under the bonnet and does not seek to find out. But who, if not physicists, should be not naïve drivers, but competent mechanics? It is no accident that the word mechanics is the keyword in the name of all areas of physics. If the profession is a "physicist," then, regardless of the field of specialization, the preference of an experimental or theoretical path, the responsibility by definition includes studying and explaining the mechanism, and not a description at the careless driver level: the wheels move the car, the car moves the wheels.

Of course, theoretical physicists sensed and understood the grotesque situation, and therefore laughed it off as best as they could. Logical circles were caused not by impaired thinking (although this is possible) but by the lack of an adequate model that describes the mechanism. In such a situation, a dead-end of a closed cycle inevitably arises. It is no different from the ancient cycle of logic, caused by incomprehension of the physical processes: there is God, and God is everything. Standard physics models use one answer to all questions: this is a field, and it has force. They could say honestly: the field is God, and it has the power to create everything and organize everything. Modern physicists try to avoid the idea of God (they are physicists!), but it is implicitly present, since all explanations come down to relying on transcendental forces with inscrutable ways. The difference between the concepts of the scientific era is often only in the words used.

Here it is appropriate to cite an excerpt from the pamphlet of two Catholic theologians Antoine Arnault and Pierre Nicole, who at the very beginning of the scientific revolution wrote with biting sarcasm: "We prefer to speculate imaginary reasons for things to be explained instead of admitting that they are unknown to us. And the manner in which we evade this recognition is quite amusing. Seeing any action, the reason of which is unknown to us, we imagine that we discovered it by adding to this action the general word "power" or "ability," which does not form any other idea in our mind except that this action has a certain reason. But we were well aware of it even before we resorted to this word. For example, everyone knows that our arteries are pulsating, that iron, being near the magnet, is attracted to it, that cassia weakens, that opium euthanizes. Anyone who is not a scientist by profession and is not ashamed of ignorance, frankly admits that he knows these phenomena, but their cause is unknown. Scientists, who cannot say this without shame, come out of the situation in a different way and pretend to discover the true reason for these actions, which consists in the fact that pulsation is inherent in the arteries, the magnet is magnetic, the cassia is laxative, and opium is a sleeping pill" (Arnault, Nicole, 1664).

They saw the mote in the eyes of others but did not notice the log in their own. The religious concept they professed explained all these phenomena using the word "power," only united everything at once in the "power of God." This manner is no less amusing and no less tragic. Although criticism of theologians can be turned on them, it is accurate in itself. The crown of Newton's concepts, who was only starting a scientific career in the year this pamphlet was published, is the theory of gravity. But it does not give anything in terms of understanding the physical essence of the phenomenon. It uses the same concept of force as a way of evading recognition of unknown causes and leaves the physical phenomenon of attraction of bodies at the mercy of the mysterious, incomprehensible, hidden, but omnipresent entity Gravity. The result was an "explanation": the apple gravitated and fell on Newton's head because of the force of gravity. The difference between the scientific method and the religious one was only a mathematical method for describing the phenomenon using special terms. But in general, Newton could easily replace "gravity" in the formula with "God's power." He did this at the end of his career, recognizing his inability to explain and calling for help from the divine power of "a Being incorporeal, living, intelligent, omnipresent" (Newton, 1704).

There is no doubt that Newton's gravitational formula gave much greater predictive power than the theologians' hypothesis of the power of God. They simply did not bother to describe the power of God with formulas. Therefore, with all the explanatory power of this hypothesis, it did not give anything in terms of prediction. Thanks to Newton's equation, it became possible to link events and phenomena into causal relationships and predict their development. At that level of knowledge, even the approximate accuracy of his linear formula was astounding and influenced the fact that in physics, a rigorous mathematical description of the "laws of forces" is stilled preferred to understanding the physical essence of phenomena. A tool (mathematical way of describing) has become a substitute for knowing the causes and mechanisms. This fully applies to the concepts of "electric force" and "charge force." In SM, a powerful mathematical apparatus allows one to describe many observed phenomena but does not explain anything.

"What is 'electric charge'? We haven't a clue" (Hotson, 2002). This truth hides behind circles of tautologies, a myriad of created entities and solemn assurances that "the origin of charge is from certain types of subatomic particles which have the property of electric charge. Electric charge gives rise to and interacts with the electromagnetic force, one of the four fundamental forces of nature. The most familiar carriers of electrical charge are the electron and proton … Current flow can be understood in two forms: as negatively charged electrons, and as positively charged electron deficiencies called holes. These charges and holes are understood in terms of quantum physics" (Wikipedia, "Electricity").

So, the charge creates the force and, at the same time, interacts with it. But the origin is undoubtedly in the particles that have this charge (or the force?). What an incredibly clear explanation. Everything, even the holes, is creating itself in this circle of life. May the force of Nature (or God?) be with it.

The practical experience of millennia suggests that there is such a phenomenon as attraction and repulsion. The ancient Greeks studied the effects of various objects attraction by amber rubbed on the wool ("electron" means amber in ancient Greek). As the experience developed, there were more and more examples of attracting and repelling at different levels of matter. We came up with new names for various forces (the tradition of animism to create spirits for each phenomenon). It is believed that bodies are carriers of charges. At the micro-level, these are particles. The model goes like this: if in a piece of matter there are more electrons than protons, then it will have a negative charge; less — a positive charge; if the same number, it will be neutral. The charge obeys the conservation law, and in an isolated system, the amount of positive charge minus the negative cannot change. The charge is discretized into integers of the elementary charge. Charges create an electric field, and if they move, they also generate a magnetic field. It is historically accepted that the direction of the current coincides with the movement of positive charges. But the current itself is considered to be the movement of charged particles regardless of their charge sign. If electrons (initially called negative) move in the conductor, it turns out that the current flows in the opposite direction to the motion of charge carriers.

The charge sign is a symbol for the external manifestation of an internal mechanism leading to interaction and movement in some direction. What is this mechanism? That is the central question. To call a charge negative or positive does not mean that the mechanism of interaction, which manifests itself in movement in a particular direction, is described and understood. Imagine that we called the movement of the clock hands to the right "positive" and to the left "negative." We do not know anything about the internal mechanism that creates such a movement, but we can use it for our purposes. We get used to the fact that our terms help us use the clock readings, but we forget that use does not mean understanding. Everything is fine as long as the clock behaves predictably. But if a malfunction occurs, then a name for the symptom is not enough. We need a diagnosis of the internal mechanism. Then we can hypothesize about causal relationships of external manifestation and internal processes. For example, we can associate the direction of movement of the internal mechanism and its parts relative to each other with the hand's movement. We might think that if the hand moves clockwise, then the insides move the same way. We disassemble the clock and are surprised to see that everything moves in different ways. If we do not understand the design of the mechanism, then it's a mystery.

Only by understanding the dynamics of the entire system and the principle of interaction of elements, we can build more reasonable hypotheses about cause-effect relationships. If we realize that the elements' internal dynamics, their frequency and phase in a coordinated interaction lead to the external manifestation of a uniform and accurate movement of the hand, then there is no mystery. With the same parameters of their phases and frequencies, these same parts and the same mechanism can move the hand in the "negative" direction. Having made such a general assumption (theoretical model), we can proceed to a detailed study and practical modeling. In this case, the quality of our understanding of the process leads to a qualitatively different level of use. We can not only use the mechanism when everything "under the bonnet" works fine, but we can repair the malfunction or even design new and better systems. We develop from an easy rider to a responsible and creative one.

But imagine that we do not understand the clock mechanism but simply say that the movement of the hand in different directions is explained by the presence of two different directions of force inside the clock mechanism. There is an apparent category mistake: the direction of external movement is equated with the direction of internal processes. These categories are related but not the same. They cannot be replaced by each other. Directions may coincide or not coincide, but without understanding the mechanism, a conclusion is impossible. But suppose that we decided to explain the internal mechanism by an external manifestation, i.e., we say that the movement of the hand in different directions defines the operation of the clock mechanism and thus determines itself. This is an even worse mistake: a cycle of logic in which what has to be explained becomes an explanation of itself.

With such a description, the errors become obvious and even ridiculous. But we are so used to them when we talk about basic physical processes that we do not notice the absurdity of the formulations within the framework of the corpuscular paradigm. Any textbook or encyclopedia will draw the interaction of charges like this:

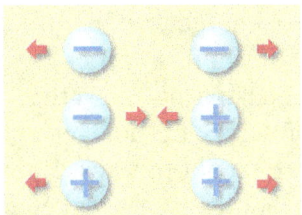

What is moving? The particles. Why do they move from each other or to each other? They have the same or different charge type. Why do they have such a charge? Because this is the direction of the particles' movement. The circle is closed. Logic at the level of a child who does not know the mechanism of the car. What moves a car? The wheels. What moves the wheels? The car.

Ask any passerby with secondary education, an electrical engineer with higher education, or even an academician of physical sciences. When asked why the charges are attracted and repelled, they will answer without hesitation: because they have different or identical signs. Well, do not ask anyone so as not to put yourself or others in an awkward position, but read the mainstream platform: "There are two types of electric charge: positive and negative ... Like charges repel each other and unlike charges attract each other" (Wikipedia, "Electric charge").

Is the explanation clear? Of course, even the child will understand. But a curious child will ask again: why are they attracted or repelled? Isn't it clear? They have different types of charges. But the child clarifies the question, as he feels the trick: how does what you call the type of charge arise? It is an uncomfortable childish question. Here, a passerby with secondary education or an academician with a higher education goes away from a curious child. Perhaps, the first one always passed difficult questions by, and the second would not have become an academician if he had not worked in the mainstream, where the answers have already been "found." All adults run away, and the child just asks the basic questions of cognition: why and how. This is not to say that adults have no explanation at all. A passerby may, of course, not philosophize craftily but simply leave. But an academician can give a whole lecture on how balls recognize each other by gender (charge sign). He can draw complex Feynman diagrams and describe them with even more complex formulas. These diagrams and equations will say that the balls have some rules of behavior, something they are allowed and forbidden. But sometimes they violate the prohibitions and misbehave.

Everything would be fine, but this is only an illusion of explanation: a description of the symptoms, which can be very vivid without revealing the internal mechanism and physical process. Such an answer allows us to describe anything but does not allow us to predict. As soon as the invented law is violated, an exception to the rule is created immediately, which becomes a new rule as the exceptions accumulate. It is a classic ad hoc/post hoc approach.

In this way, the mainstream describes not only gluons, bosons, etc., which are far from the prose of life, but also electrical phenomena close to every passerby. In the "billiard ball" paradigm of electricity, in various types of matter different particles and even holes can be charge carriers. The latter are introduced as quasiparticles with a positive charge to explain "hole conductivity" when there are no electrons but electricity flows. Remember the definition of electricity? It is the movement of particles and quasiparticles. At the school curriculum level, they talk about electrons because children are not yet ready to "digest" the concepts of abstract holes when they are taught physics. But university students have to shut up and calculate how virtual holes create real energies.

There is even a "band theory" in solid-state physics that "describes the range of energy levels that electrons may have within it, as well as the ranges of energy that they may not have (called band gaps or forbidden bands) (Wikipedia, "Electronic band structure"). Quantum mechanics allows and forbids its particles (electrons) to be in certain states. According to the Bohr model, electrons can only be in one orbit, and their energy takes strictly discrete values. We have already examined the miracles associated with this model, but the players of the "Particle" team have an inexhaustible supply of miracles.

"Imagine a row of people seated in an auditorium, where there are no spare chairs. Someone in the middle of the row wants to leave, so he jumps over the back of the seat into another row and walks out. The empty row is analogous to the conduction band, and the person walking out is analogous to a conduction electron. Now imagine someone else comes along and wants to sit down. The empty row has a poor view, so he does not want to sit there. Instead, a person in the crowded row moves into the empty seat the first person left behind. The empty seat moves one spot closer to the edge and the person waiting to sit down. The next person follows, and the next, et cetera. One could say that the empty seat moves towards the edge of the row. Once the empty seat reaches the edge, the new person can sit down. In the process everyone in the row has moved along. If those people were negatively charged (like electrons), this movement would constitute conduction. If the seats themselves were positively charged, then only the vacant seat would be positive. This is a very simple model of how hole conduction works … In reality, … the hole is not localizable to a single position as described in the previous example. Rather, the positive charge which represents the hole spans an area in the crystal lattice covering many hundreds of unit cells. This is equivalent to being unable to tell which broken bond corresponds to the "missing" electron. Conduction band electrons are similarly delocalized" (Wikipedia, "Electron Hole").

A simple model, isn't it? The only problem is that it just postulates the rules of seat occupation by the electrons. There is no answer to the main question about the mechanism. Why would electrons choose certain seats? How do they define a seat which is so spread out that it covers hundreds of seats? Without an understanding of the mechanism "under the hood", the rule becomes an axiom, in which one can only believe or not believe. But the problems begin with the first check of the engine. For example, the "band model" does not agree with the phenomena of superconductivity, ferromagnetism, and even conductivity in semiconductors, dielectrics, and metals in a particular state.

As a result, the quantum theory of semiconductors cannot accurately calculate any constant of a semiconductor. It is impossible to predict, knowing the structure and composition of a substance, whether it will go into a superconducting state with decreasing temperature. It is impossible to even approximately calculate the magnetization, heat capacity, electrical conductivity, and other macroscopic values based on the known structure of the crystal, the electron shells of atoms in the crystal, and other parameters of the microworld for strongly magnetic substances (ferromagnets, antiferromagnets, and ferrimagnets). In most semiconductors, the magnitude and temperature dependence of thermal electromotive forces diverge from the theory's predictions.

Let's take another classic description of the motion of electrons in metals called the Drude model. Three years after Thomson announced that the electron was a particle, Paul Drude came up with a model that "provides a very good explanation of DC and AC conductivity in metals, the Hall effect and the magnetoresistance in metals near room temperature" (Wikipedia, "Drude Model"). It is good that it provides a good explanation. Let's see how. In this model, it is believed that electrons, like gas atoms in kinetic theory, are identical solid spheres that move in straight lines until they collide with each other. It is assumed that the duration of an individual collision is negligible and between the molecules no forces other than those arising at the time of the collision act. Electrons, for some reason, move with a constant speed in a straight line. These colliding solid balls change speed, but the speed after a collision is not related to the speed before a collision and is random. Such is the "good explanation" within the model.

How can a body's speed after a collision not be related in any way to the speed before it? What does random direction after a collision mean? The final vector depends on the sum of the input vectors and is not something random, but is a result of the interaction of different parameters. If the balls fly randomly and regardless of the cue and other balls, there is no way to play a billiard game. But maybe it's about another game? So it is: "The model ... assumes that the microscopic behavior of electrons in a solid ... looks much like a pinball machine, with a sea of constantly jittering electrons bouncing and re-bouncing off heavier, relatively immobile positive ions":

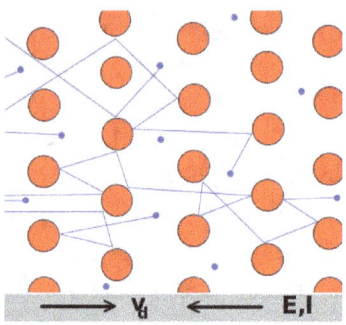

Wikipedia, "Drude model"

Well, let it be this way: the balls fly in the matter, somehow collide and fly apart. But what gives them direction and force of movement? Here is the same circle of logic: the direction creates an electric field, and the electric force provides the force. Again, magical powers and fields. Well, let there be balls, forces and fields. Does the model fit empirical data? It is believed that the two most important consequences of the model are the formulas of electronic motion and a linear dependence of the current density and electric field, where these parameters are connected through the following variables: elementary charge, electron concentration (number of balls), ball mass and relaxation time (ball free travel time between collisions):

$$J = \left(\frac{nq^2\tau}{m}\right)E$$

"The latter expression is particularly important because it explains in semi-quantitative terms why Ohm's law, one of the most ubiquitous relationships in all of electromagnetism, should hold" (Wikipedia, "Drude model").

It is a typical example of a category error: description is taken for an explanation. The formula simply reflects how the model describes the relationship of parameters. But these variables do not have any explanation within SM. Of course, if we do not take their "semiquantitative" nature for an answer. Sometimes they say that the model uses such unclear parameters because it is purely phenomenological. Typically, phenomenological models include some descriptive concepts that do not consider the observed phenomena' internal mechanisms. By definition, phenomenology only describes external manifestations, which separates it from scientific theory and experiment, designed to confirm or refute the hypothesis. Often these descriptions include the results of observations that cannot yet be explained by existing theories.

The theory of valency in chemistry can also be called phenomenological since it is utterly helpless in describing the physical mechanism underlying the combination of elements. The Ginzburg-Landau theory, which describes superconductivity but does not explain its mechanism, is also of the same kind. The classical Faraday's theory of electricity is also not about the mechanism but about the description of manifestations. But at least the description should coincide with the phenomena and not turn into a hallucinatory "castle in the air," detached from physical reality.

We see that the semi-quantitative variables in equations, "explaining" the physical phenomenon of electricity, are inexplicable and mysterious, and the model does not correspond to the empirical phenomena. It can be considered only a simplified, approximate linear description only for idealized conditions (uniformity and constancy). As soon as we get to reality, the model breaks up. It cannot adequately describe physical phenomena in strong magnetic fields. Even for weak magnetic fields, the Drude theory loses its applicability due to phenomena related to interference, for example, weak localization, the Aharonov-Bohm effect, fluctuations of conductance, hopping conductivity, superconductivity, etc. But the model, which does not adequately reflect reality, is considered a classic description and even an explanation of Ohm's law.

Here it is worth briefly dwelling on this law, which every physicist, electrician and almost every schoolchild knows. It establishes a proportional relationship between the source's electromotive force (electrical voltage), the current flowing in the conductor, and the resistance of this medium. Georg Ohm's original formula was different, but we give the standard for textbooks:

$$I = \frac{U}{R}$$

In words: the strength of the current in the circuit section is directly proportional to the voltage and inversely proportional to this circuit section's electrical resistance.

Let's try to figure out the variables in this formula. Ohm was not the first to experiment with electricity and not the first to discover the relationship between flow velocity and voltage. Half a century before Ohm in 1781, Henry Cavendish studied the current strength on his skin: he measured the power of the electric shock with his body at different "speeds" and the "electrification level" (now called voltage). Pay attention to his terminology. It accurately reflects the essence of the process's physics: the speed of energy propagation and the energy level.

Let's read the standard definitions of the variables specified in Ohm's law and the basic concepts associated with it.

"An electric current is the rate of flow of electric charge past a point or region" (Wikipedia, "Electric current"). "Voltage, electric potential difference, electric pressure or electric tension is the difference in electric potential between two points" (Wikipedia, "Voltage"). "The electrical resistance of an object is a measure of its opposition to the flow of electric current. The reciprocal quantity is electrical conductance, and is the ease with which an electric current passes" (Wikipedia, "Electrical resistance and conductance").

At the phenomenological level of description, everything is quite simple and intuitive. It is based on a hydrodynamic analogy. Current as the water flow, voltage as the water pressure between two points, resistance as the friction of the water channel. The problem starts when an explanation is needed. If it is a current, then what flows? If charged particles flow, then what is a charge? If the charge is a manifestation of interaction, then what kind of interaction? If there is an interaction, is it direct or indirect? If mediated, then what is the intermediary? If direct, then what is the mechanism? And if there is an intermediary, then what is the mechanism of interaction with the intermediary? Everything rests on the mechanism of interaction and its laws.

Ohm described the relationship of parameters but did not say anything about the mechanism. Drude proposed the model that proceeded from the notion of an electron as a particle and took the formula from the model of an ideal gas. The electrons were considered perfect balls in an ideally isotropic medium (the so-called "electron gas"). Later the model was updated and finalized in the framework of Quantum Mechanics, but the paradigm remained the same. How did it explain Ohm's law? Electrons fly in their zigzags in the "pinball machine" of the conductor but generally shift in the opposite direction to the electric field. Why?

According to the Pauli principle, two electrons cannot be in one quantum state: they must differ by at least one quantum number. We have already examined the axiomaticity of this law, which can also be called phenomenological, since it merely states and describes but does not explain the mechanism. Attempts to describe the mechanism within the "pinball" model led to various magical tricks and transformations. An electron becomes a spirit that can be everywhere and nowhere, simultaneously at different poles of an atom and different phases of its orbit, and it is a traveler in time. It also has the magic ability to push other balls with the same quantum number so that they do not take the same place with it. How the electrons detect the quantum number and other signs of each other, decide whom to allow into the home orbit and whom to expel, remains "behind the scenes" of the theory because it would require a description of the magic interaction of these spirits. They simply prefer a statement: the same balls cannot be together. Period. And how they identify each other is their business.

In this model, the system's energy is made up of the energies of the particles, which are simply added, filling the states with increasing energy (literally filling the atom's orbitals with electrons). The energy of the system is called the "Fermi energy" and is one of the central concepts of solid-state physics. It is the energy difference between the highest and lowest occupied single-particle states in a quantum system. What is the physical meaning of the Fermi energy level? It is the probability of detecting a particle at the Fermi level.

So, there is a kind of "empty system" (for example, an orbital) being filled by particles: like a wire filled with birds sitting on it one at a time. These birds cannot sit at the same place and share the wire among themselves: only a certain number and only regular birds can fit on it. This wire's physical meaning is the probability of finding the right bird on it, and this probability is 1/2 at any temperature except absolute zero. Thus, the electron is either there, or it is not there. Here is the physics of the process: the likelihood that everyone sits in his places or flies in the void.

How does electrical conductivity occur? The answer is simple: the field applies its force to these "birds on wires." Here we will not even question what kind of field this is and where it comes from. We concentrate on the birds on a wire. So, the wind (field) blew, and the birds (electrons) moved where it was blowing, but only on the condition that there were empty spaces called momentum states. As a result, the more vacant seats, the easier it is to move around, which means that the substance is more electrically conductive. The insulator has very few or no seats at all.

Of course, the substance is not represented as a one-dimensional wire but as a three-dimensional lattice, but the essence does not change: birds move in the void and sit on different wires with empty seats. The wires themselves turn out to be levels requiring filling or emptying. It is a system with gaps and levels, the topology of which is directly related to the electrical conductivity.

Electron birds must be free to move in the void, and for this, they must be brought to the "excited state." What do they need for this? They must swallow photons. Depending on the availability of free space and electrons' ability to eat photons flying there, we have either a conductor, a semiconductor, or an insulator. That's the whole mechanism. Briefly, it is like this: the movement of electrons through matter in the void between its lattices. Choose which metaphor suits you: either pinball or bird flights. The essence does not change. The quantum model and the Drude model are about the same thing: particles roll and fly in the void. What drives them? Force and field.

Both models indicate that the conductivity is proportional to the number of free electrons (population density). The speed (current strength) is equal to the rate of movement of these particles, which depends on the power of the external electric field and the resistance of the conductor. Both models use auxiliary variables. The difference is that the quantum version uses additional auxiliary variables "Fermi level," "Fermi speed," "Fermi sphere," "Fermi surface," and others. The Drude model spoke of all free electrons, and the SM speaks of special electrons of the Fermi level with the Fermi velocity. This is considered a more accurate, adequate and rational explanation than the classical model. Still, both models face the same problems when the question arises about many phenomena of conductivity.

But the main problem remains the disclosure of the physical mechanism of interaction. Both talk about some entity "field" that drives loose balls in the pinball machine of matter. They collide and fly apart, fly randomly in the void between the "planets" of atoms, hanging in the void in their orbits, or take the directional motion under the influence of the field as a "holy power" needed to explain everything by itself, including itself.

Faraday and Maxwell worked at a time when ether was not yet banned. It was a time of physics using math to describe energy interactions in the medium. But then came mathematical physics describing phantoms in the void. Maxwell wrote: "Faraday in his mind's eye, saw lines of force traversing all space where mathematicians saw centers of force attracting at a distance. Faraday saw a medium where they saw nothing but distance. Faraday sought the seat of the phenomena in real actions going on in the medium, they were satisfied that they had found it in a power of action at a distance impressed on the electric fluids ... In explaining the electromagnetic force by means of a state of stress in a medium, we are only following out the conception of Faraday ... We have asserted nothing as yet with respect to the mode in which this state of stress is originated and maintained in the medium. We have merely shown that it is possible to conceive the mutual action of electric currents to depend on a particular kind of stress in the surrounding medium, instead of being a direct and immediate action at a distance" (Maxwell, 1873).

Classical electrodynamics was based on the wave theory, and waves propagate in the medium. Physicists of the 19th century still adhered to the logic of physical causation. If interaction extends over a distance, it means there is a physical environment for its propagation. But the corpuscular model and the abolition of the ether by Einstein prevented drawing conclusions within the framework of physical logic. Disciples of the new faith had to write all the time about some non-material spirits.

"In quantum theory, the electromagnetic field behaves exactly as an assembly of arbitrarily many massless particles. The number of particles of a given momentum and energy just corresponds to the energy level of the corresponding electromagnetic oscillator" (Taylor, 2001). Thus, some spirits without mass but with energy form the total energy of the field. We forget about the equivalence of mass and energy, although it is one of the basic laws of physics, regardless of the interpretation of the generating mechanism. We forget about physical causality and logic. As a result, we get a certain arbitrary number of incorporeal spirits that create the real energy of electromagnetic oscillations.

But common sense suggests the inevitable questions. "Since Einstein banished the aether, however, the question has been "what is waving?" The photon has no rest mass, and contains no charges — so it violates our kinetic definition of energy" (Hotson, 2002). In principle, it violates all concepts about physical reality. It is an incorporeal angel of light, flying in the void with absolute and constant speed, and carrying on its wings the real electromagnetic interaction observed between the material bodies of the world. How does energy transfer take place? Some electrons emit photons, while others swallow them. How do they do it? Here all the answers boil down to one: spirits live according to their own laws, which are unknown to us. Do not forget about the most important law of spirits: they fly in the void. Literally: "In the special theory of relativity, the region of space without matter and without an electric field appears completely empty, i.e., it cannot be characterized by any physical quantities" (Einstein, 1918). How do photons move in the void? Easily, as nothing hinders their flight on their angelic wings.

Amazingly it is a summary of the theory of electromagnetic interactions in its modern form in mainstream theoretical physics. At the same time, the practical use of electricity is based on the models of Faraday, Maxwell, Ohm, and other classics who thought in terms of real energy interactions in the medium. What they lacked was an idea about the mechanism of such interactions. Unfortunately, at the beginning of the 20th century, the choice was made in the direction of using phantoms in the void to describe real interactions in the medium.

To return to real physics in our theories, we need a paradigm shift that will cancel a huge number of phantoms with their specialization in different areas of fundamental interactions and create a unified model that describes the laws of the same process at different levels. This is what the *Theory of Energy Harmony* (TEH) is all about.

To change the paradigm, we cancel the basic assumption of the corpuscular model of the world about the existence of elementary particles, as some indivisible "bricks of the universe." We proceed with the assumption that what exists are energy oscillations with certain amplitude-frequency parameters and phase portraits that form structures that we perceive as forms of matter, including microcosmic ones that we call particles. In this case, energy interaction turns out to be a process of frequency and phase coupling and uncoupling (synchronization and desynchronization).

We have looked into the details of this mechanism in the previous part of the study. Here we will be formulating hypotheses concerning all fundamental interactions in general and each one in particular.

Hypothesis:

All types of fundamental interactions (strong nuclear, weak nuclear, electromagnetic, gravitational) with their corresponding phenomena of attraction, repulsion and balanced state are manifestations of the same mechanism of frequency-phase coupling (synchronization) of energy oscillations at different frequency levels. These oscillations propagate as waves in a continuous, all-encompassing and active energy medium. The movement and transfer of energy are carried out in wavefronts with various configurations (running, standing, flat, longitudinal, transverse, longitudinally transverse, bending, spherical, and others) creating different interaction propagation vectors and their relationships manifested in the observed patterns of interactions.

CHAPTER 3

THE HYDRODYNAMIC ELECTROMAGNETISM

In order to obtain physical ideas, we must make ourselves familiar with the existence of physical analogies.

James Maxwell

The idea of any analogical model is simple: to describe a target system that is under investigation by comparing it with the analogous system that manifests itself by similar phenomena and has been studied more. But sometimes using the analogy is tricky as both systems can still be an enigma. In this case, we need to go deeper and look for the mechanism "under the bonnet" of both systems. If it turns out to be the same, the analogy is fruitful. Otherwise, we should change the analogy for the one that will lead us to the understanding of both systems.

To illustrate this subtle issue, let's take an article by one of the teachers of physics called "The Role of Models and Analogies in the Electromagnetic Theory: A Historical Case Study" (Silva, 2006). The abstract describes the state of affairs in physics departments of universities: "Despite its great importance, many students and even their teachers still cannot recognize the relevance of models to build up physical knowledge and are unable to develop qualitative explanations for mathematical expressions that exist within physics. Thus, it is not a surprise that analogies play an important role in science education, since students' construction of mental models of abstract phenomena need to be rooted in some existing or previous experience in order to interpret more complex ideas" (Ibid).

The author notes that there are "two kinds of analogies: the formal analogy and the material analogy. In the first case, the same axiomatic and deductive relations associate both subjects and objects of similar systems. In this case, these relations are described through similar equations. For instance, a pendulum and an oscillating electric circuit are formally analogous since both systems can be described through the same differential equation. The existence of a material similarity between both systems is not necessary. When material analogies are taken into account, there is a physical similarity between the systems; as an example, one can take the kinetic theory of gases that considers a gas as being a set of tiny spheres. Gas molecules may be conceptualized as a collection of billiard balls randomly moving and hitting one another, even though gas molecules and billiard balls are not quite identical" (Ibid).

Describing the history of electromagnetic studies, the author insists that the pioneers "studied physical phenomena that were completely unlike, such as heat flow problems, electrostatic attraction and gravitational attraction," though they "realized that such phenomena could be described through equations of the same kind by simply attributing the right meanings to each symbol, in each equation" (Ibid).

It is striking that the author calls the billiard ball analogy, which leads to numerous contradictions with the described physical process, a material one. But the analogy between two oscillatory systems turns out to be formal. Everything is turned upside down. The processes that are fundamentally similar and differ only in detail are called "completely unlike." But it is exactly their physical similarity that leads to the fact that they can be described by equations of the same kind. It speaks not of the formality of the analogy but of its materiality, of the presence of a physical basis for the associative assignment of phenomena to one class. Their details may vary, but the essence is the same.

But where does this error of attribution of similar phenomena to different classes come from? We are not surprised to see that it comes from the same old question about the medium. The author reveals the problem: "William Thomson, James C. Maxwell and others developed models and analogies to explain both electric and magnetic phenomena based on the existence of ether. The equations and physical concepts that were drawn through the analogical method are still used and taught until today, despite the fact that we do not believe that a material medium such as ether pervades space" (Ibid).

Teachers use analogies with processes in a material medium, but are forced to listen to the message of the prophet Einstein about the abolition of the ether, constantly and persistently broadcast by the preachers of the new religion to their flock. The author's use of the verb "believe" is not accidental. It is not from a scientific, but from a religious dictionary. Science is not about beliefs, but about creating the most probable model of the world and testing it against reality. A person who thinks in terms of "I believe/I don't believe" is outside the scientific discourse. Unfortunately, the author of the above article does not even notice that the phrase "we do not believe" puts the entire community of physicists who hold such views outside of physics as a science. The religious approach in theoretical physics became so common in the 20th century that it passed without a hitch into the 21st century and continues to be mainstream. This is why the verb "believe" is a normal part of an author's active vocabulary when writing an article for a scientific journal.

Not believing in something is also an act of faith, since it presupposes unconditional acceptance of some idea as truth. The phrase "we do not believe that a material medium exists" means "we believe that emptiness exists." This is exactly what Einstein said when, instead of the ether, he proposed empty space, which "cannot be characterized by any physical quantities" (Einstein 1918). Logic inevitably leads to the fact that a model based on emptiness begins to produce immaterial entities living in it.

The author writes: "It is not sufficient that students are able to deal with the mathematical aspects of a theory. In order to be able to build mental models of the electromagnetic phenomena, understanding the physical concepts involved in the mathematical formulae is a necessary condition" (Silva, 2006). This is the crux of the problem. Physics must contain a physical meaning, described in words and equations. But emptiness, which cannot be characterized by any physical quantities, has no physical meaning. This obvious conclusion creates a powerful cognitive dissonance that underlies the main models of theoretical physics created in the last century and still leading in this century.

Einstein himself tried to overcome it all his life and either abolished the ether, then revived it, then created a surrogate environment (space-time fabric), then canceled it (more on this in the next chapter). Meanwhile, practical physicists did not pay attention to these quirks but used the classical model of Faraday and Maxwell, which was based on the existence of a medium. With its help they created the technologies on which all modern civilization is based.

The author of the article states: "Teachers very often use some medium like ether to explain the immaterial processes that take place in the "empty space", however it occurs in a subliminal and unconscious way" (Ibid). Everything is exactly the opposite. When an explanation of material processes is required, teachers try to talk sense about the physical environment. This is a conscious act. But if they talk about "immaterial processes in the empty space," it is the unconscious repetition of mantras from the messages of the prophets that do not make sense.

Let's try to remain mindful and use material analogies that have physical meaning. In some cases, it will come easy, but in others, it will require a struggle with habitual perceptions based on meaningless dogmas. Let's start simple and look again at the Ohm's law equation: current depends on voltage and resistance. If we take the hydrodynamic analogy, even a secondary school student can understand it. There is a flow, and its speed depends on the height difference (or another factor that creates a difference in potential of energy levels) and on the resistance of the channel and other obstacles in it. But such a simple and clear analogy requires the existence of a medium. The analogy does not work for the leading physics theory based on the void. But in schools, they prefer to give the old, classic version, in which electrodynamics go along with hydrodynamics. Children love fairy tales but at the lesson of natural sciences they expect the teacher to make physical sense. If he starts to talk about miracles in the void, they may ask awkward questions for which the teacher does not have answers.

Actually, for describing electrical conductivity, Ohm used the work of Joseph Fourier on thermal conductivity. Thus, he used hydrodynamic and thermal analogies together. It is not a coincidence but the direct result of observing the same manifestations of energy transfer. Ohm's law is the electrical analog of Fourier's law of thermal conductivity at various temperatures. If we look at the variables in the equations deeper, we understand that they actually represent wave parameters that depend on the source and the medium of propagation. The oscillations at the source determine the initial wave characteristics. The oscillatory conditions of the medium determine the resulting energy transfer. Ohm's law calls these variables current strength, voltage, and resistance. Fourier's law calls them the heat flux, temperature gradient, and thermal resistance (or conductivity). But they are all about waves in the medium. Our habit of giving varying labels to the same energy processes stems from a lack of understanding of the underlying mutual mechanism.

It is interesting to note that when it comes to energy propagation through some form of matter it is called conduction and when the same process happens between bodies it is called radiation. If energy is transferred through some wire, it is called electricity. If the same energy goes from the Sun to the Earth, it is called electromagnetic radiation (light). If energy is transferred within some body, it is called heat flow. If the same energy goes from the Sun to the Earth, it is called thermal radiation.

It is more interesting to note that this difference in terms is interpreted as a conceptual difference: "In solids, conduction is mediated by the combination of vibrations and collisions of molecules, propagation and collisions of phonons, and diffusion and collisions of free electrons. In gases and liquids, conduction is due to the collisions and diffusion of molecules during their random motion. Photons in this context do not collide with one another, and so heat transport by electromagnetic radiation is conceptually distinct from heat conduction by microscopic diffusion and collisions of material particles and phonons" (Wikipedia "Thermal conduction").

So, conduction is about particles and quasiparticles movement. Radiation is about particles movement. What is the conceptual difference? At first glance, there is none. But we must remember that according to the mainstream paradigm photons as carriers of electromagnetic interaction are angels without mass that fly through the void and carry energy on their wings without any resistance. How would the void support any energy transfer if not for this magic flight?

But if we get back from fairy tales to the physical reality, we understand that resistance is the other side of the conductivity. The medium can be more or less conductive (less or more resistive). The void has zero conductance as there is nothing in it to conduct anything. We can put it the other way: the void has infinite resistance. If something passes through the void it must have the magical property of angels that do not need any support for their wings. They just go straight from the divine source without hindrance and without any interaction with one another. What is the result? "Electromagnetic radiation, including visible light, will propagate indefinitely in vacuum" (Wikipedia "Thermal radiation"). Why do we call it radiation then? Let's just call it indefinite angels' propagation.

The corpuscular paradigm attributes various names to these angels. Some of them are called fundamental particles, some are virtual and quasi, and some are just holes. This is how the same patterns of real interactions at different amplitude-frequency levels of the energy vibrations are transformed into the names of phantom entities. But the paradigm insists that they are real as they can be counted. Thus, the discreteness of our measurement of continuous energy (quantization) turns into the discreteness of this energy (quanta as particles). Here is a typical example of such an objectification error: "Phonons can be thought of as quantized sound waves, similar to photons as quantized light waves. However, photons are fundamental particles that can be individually detected, whereas phonons, being quasiparticles, are an emergent phenomenon" (Wikipedia "Phonon").

We have considered this "individual particle detection" that is actually wave measurement technology in the previous part of the study. Here we will just note that the author of the above statement expresses mutually exclusive ideas within two sentences: photons are quantized waves and photons are fundamental particles. The first one reflects reality. The second one reflects the objectification error which is at the heart of the corpuscular paradigm that creates phantom particles (entities) out of quantization (measurement process). We have also dealt with this issue earlier, so we will just stress that the electrons, positrons, photons, phonons, and holes are not particles, antiparticles, and quasiparticles, but energy oscillations with a certain phase trajectory and amplitude-frequency characteristics that result in a stable configuration of their phase portrait which can be measured as whole. This produces the effect of a "detection of a particle."

Of course, we can call this phase portrait by any name we like, but if we take it for a fundamental piece of matter (particle) we inevitably face contradictions with reality when these indivisible discrete things start behaving like continuous and spreading waves. We also face an insurmountable difficulty when we try to explain the interaction of these entities: if they are discrete parts with nothing in between, their interaction becomes "spooky action at a distance" or the flights of interaction carriers in the void. Both versions are outside of the scientific discourse and do not have any physical meaning, though they are at the heart of Quantum Mechanics and Standard Model which are called physical theories.

The reality of electromagnetic or thermal or any other energy interactions is about the propagation of waves in a medium. This medium can effectively synchronize with incoming oscillations thus facilitating wave propagation, and we can call it a conductor. If a medium is poorly synchronizing with this particular frequency range of energy vibrations, we can call it an insulator. There is no conceptual difference between conduction within a body and radiation from one body to another: it is all about wave transfer of energy. The only difference is in the parameters of the propagation medium that determines the observed effects.

If we flip Ohm's equation, it says that the conductor's resistance increases in direct proportion to the energy applied to it. But this does not correspond to reality. The medium's resistivity (or conductivity) is obviously related to the amount of transmitted energy but it depends on the physical parameters of the medium itself. This is reflected in practice: ensuring efficient transmission of electricity through different conductors is the main task of power line technologies. They are all about looking for a medium that makes waves propagate better and not about finding how to push particles through the wires. But a typical electrician will say without hesitation that electricity is the movement of charged particles. This is how the corpuscular dogma works: we work with waves but think of phantom particles. In reality, what moves from one point to another in the electric line is the oscillations (waves) but not parts of the material (particles).

Ohm's law describes quite well the behavior of conductors at low "electric field strength," which means actually low amplitudes and frequencies of oscillations. At other levels, the discrepancies between the equation and reality begin to go beyond the acceptable approximate accuracy. The direct proportionality no longer works even in a conventional incandescent lamp as its current-voltage characteristic is not linear. The same goes for fluorescent gas lamps and even more so for semiconductors (diodes, transistors). Solid and liquid electrolytes, solutions, melts, gases, and other materials do not obey linear equations at all. Similarly, the basic thermal conductivity formula that idealizes the conductor as an isotropic and homogeneous medium only works approximately with specific parameters and does not work at high-frequency levels (usually called high temperature).

What Ohm's and Fourier's laws say in math terms can be expressed in words: in an ideal isotropic medium, the speed of energy propagation is directly proportional to the amount of energy and inversely proportional to the dissipation of this energy in the medium. But in a real medium wave propagation is more complicated and simple linear equations are just rough approximations. If we look from a wave perspective, all deviations from a simple proportionality can be explained: the dependence of the group wave velocity on the source parameters and propagation medium is a nonlinear function. The degree of energy dissipation is determined by the degree of coherence of the wave itself and the coherence of the medium. The resistivity of the medium is a relative concept with a nonlinear interplay of many parameters.

By the way, Fourier went further than Ohm and tried to describe the nonlinearity of the process using trigonometric functions. This revolutionary approach was not appreciated. For example, Lagrange and Laplace criticized Fourier's work for using complex functions to describe things that seemed to work in a simple way. The real significance of his method of analysis and transforms became apparent only in the second half of the 20th century. It became a powerful tool for the mathematical description of oscillations from the macro- to microcosmic level.

Electric current and electromagnetic radiation are not a movement of particles as standard definitions postulate. There is no movement of matter within matter to transfer energy in the case of electrical conduction. There is no movement of massless photons in the void in case of radiation. In both cases, what moves is the wavefront in an oscillatory medium. The parameters of the oscillatory source and the oscillatory medium define the parameters of wave propagation. In both cases, the Huygens-Fresnel principle works, according to which each element of the wavefront is a source of secondary disturbance generating its own waves. The result is determined by the interference of all waves. The configuration of these waves' fronts and vectors can be different and determine the resulting interaction patterns. Such configurations were observed by Faraday in his classical experiments with metal sawdust. However, they are not lines of force of electric and magnetic fields as they are called within the mainstream theories. These are interference patterns of waves in an all-encompassing energy environment.

We can call it by any name but we should not think of it as some kind of a separate entity that acts on matter and produces this or that result. Energy is the only irreducible and fundamental entity. All forms of matter and phenomena of their interactions are processes within this entity. There are no separate fields for each kind of interaction. There are only different amplitude-frequency levels and various phase trajectories within one entity. That is why we invariably encounter similar wave structures when we register manifestations of interactions at different levels.

The only difference between electrical conductance within some type of matter and radiation between various forms of matter is the properties of the medium. For example, an electrical wire made of a certain material will have specific conductivity parameters and work as a waveguide for waves sent from the source of the current. If it is not insulated the waves will spread outside and there will be a loss of directional energy transfer. Insulation makes the guiding properties of the channel better. Any medium is an oscillatory system that is just more or less conductive for this or that kind of wave with specific parameters.

Hypothesis:

Conductivity means the ability of a medium to synchronize with external oscillations and pass energy by the mechanism of frequency-phase coupling. As it is a universal mechanism, there is no conceptual difference between any kind of observed energy interactions.

The same goes for the phenomenon of superconductivity. Ohm's law does not work for it as it means zero resistivity of a medium. Once again, in an attempt to explain QM tries to apply its rules of the game: "Superconductivity is a set of physical properties observed in certain materials where electrical resistance vanishes ... An electric current through a loop of superconducting wire can persist indefinitely with no power source ... Superconductivity is a phenomenon which can only be explained by quantum mechanics" (Wikipedia "Superconductivity"). It is a typical panegyric to the quantum model. Let's see how it explains the phenomena. Knowing all the tricks of the quantum world that we have looked at in the previous volume, we can predict that it will be full of magic and devoid of physical meaning.

One hundred years ago the very idea of superconductivity sounded like something from the realm of fiction, and the term even had a connotation of sarcasm by analogy with the idea of a perpetual motion machine. Interestingly, superconductivity was discovered by chance, just as the "sympathy" of the pendulums, and seemed like a miracle. Further experiments showed that the current can indeed circulate through the superconductor for years without energy loss, i.e., medium resistance is zero. As initially superconductivity was discovered at temperatures close to absolute zero, the concept remained centered on the idea of what happens in these conditions.

The classical electrodynamics did not predict such an effect, so it needed a new model. Before the rule was simple: particles move randomly; when the field is applied it gives them direction and they march where it shows; unfortunately, their march is not perfect and they collide with each other and obstacles provided by other particles in the conductor; it means that it resists their marching and the energy is lost. With superconductivity, electrons suddenly cease to collide and do not lose energy.

It took a long time since the discovery of the effect in 1911 to come up with some kind of a new rule for particles: "The complete microscopic theory of superconductivity was finally proposed in 1957 by Bardeen, Cooper and Schriefer. This BCS theory explained the superconducting current as a superfluid of Cooper pairs of electrons interacting through phonons' exchange. For this work, the authors were awarded the Nobel Prize in 1972" (Ibid).

Why do electrons with negative charges that should repel form each other form pairs and go on the superfluid parade without any hindrance? There is no answer within this "complete theory." It is just a standard trick: if it looks like the game will be lost, just invent new rules, and voila — here is the "explanation" and the prize for it.

Here are new rules: "Although Cooper pairing is a quantum effect, the reason for the pairing can be seen from a simplified classical explanation. An electron in metal normally behaves as a free particle. The electron is repelled from other electrons due to their negative charge, but it also attracts the positive ions that make up the rigid lattice of the metal. This attraction distorts the ion lattice, moving the ions slightly toward the electron, increasing the positive charge density of the lattice in the vicinity. This positive charge can attract other electrons. At long distances, this attraction between electrons due to the displaced ions can overcome the electrons' repulsion due to their negative charge, and cause them to pair up ... Unlike electrons, multiple Cooper pairs are allowed to be in the same quantum state, which is responsible for the phenomena of superconductivity" (Ibid).

Do you see how everything is tidy now? Elementary indivisible particles with a negative charge receive some positive charge, overcome mutual aversion and pair up. How does this transition happen? By a miracle. To be precise, by a change of the rule by the White Queen of the quantum Wonderland: electrons are now allowed to be in the same quantum state. Forget any principles, be it Pauli principle or any other. SM has the universal solution: invent a special spirit that will help the electrons break the ban and create couples.

It is not surprising that the very concept of a phonon was initially introduced as a quantum of vibration (discretization of a continuous process). In 1932, Igor Tamm introduced it for the convenience of mathematical description, since considering the structure of a solid matter in the form of vibrations of individual atoms led to trillions of unsolvable differential equations. But when the time came, the virtual rabbit jumped out of a magician's hat and the audience was introduced to the new phantom particle. According to the tradition of objectification error, the phonon-quantum has turned from a discrete measurement into a virtual phonon-particle. It got its place in the hierarchy of spirits of SM as a special agent responsible for thermal conductivity, scattering in solids, and superconductivity.

The Nobel Committee put an approval stamp, and everyone cheered the "Particle" team for the virtual goal. However, a problem arose: the phonon mechanism did not explain superconductivity at high temperatures. What was proposed? They said that it works for conventional superconductors, but it cannot be applied to unconventional. This is the essence of the explanation that is considered "complete theory." The truth is that the model is not adequate for the phenomena it pretends to explain. The other truth is that for real explanation we do not need any magical quantum properties of phantom particles but just the physical properties of a material medium.

Nobel Committee issues prizes not only for theoretical models. In 2001, physicists Eric Cornell, Wolfgang Ketterl, and Karl Wiemann received the Nobel Prize for creating a state of matter on the verge of absolute zero. A press release from the Royal Swedish Academy of Sciences said that atoms sang in unison. For many physicists, this sounds like a metaphor. But from the TEH perspective, it is a physical analogy. If we consider atoms as systems of oscillations, singing in unison means a coherent state of sync at a harmonious ratio. TEH predicts that all conductive effects are about the synchronization of oscillations in a medium with incoming waves and a transfer of energy by the mechanism of frequency-phase coupling.

What happens at ultra-low temperatures? We should remember that temperature is just a measurement parameter that reflects the kinetic energy of vibrations within the material. Even if we model this material as a set of particles, we will talk about vibrations. But when it comes to an explanation of a mechanism of conduction with the help of "balls" vibrating in the void, we face contradictions of all sorts and need to come up with new rules each time reality produces surprises. The only reason for surprises is that the corpuscular model does not have an explanatory and, consequently, a predictive power.

Alternatively, TEH does not have a problem with an explanation and predicts any level of conductivity. Low-temperature superconductivity means that a conductor as a system of interacting oscillators establishes a highly coherent structure with phase portrait parameters that allow the establishment of effective synchronization with the incoming waves. Recall the Arnold tongues: the closer to the optimal harmonious ratio of the interacting oscillations' parameters, the greater the frequency-phase coupling, and the less energy required to maintain the synchronization state. The medium acquires the features of a superconductor, which physically means the transfer of energy waves without loss as there is no additional parameter matching. The whole system becomes a single chorus, singing in unison. It is not a poetic metaphor, but a physical reality: vibrations of energy interact with other vibrations in perfect harmony without resistance, i.e., without changing the amplitude-frequency characteristics and phase portrait of the waves passing through the conducting medium. Is it a coincidence that outer space temperature is at the absolute zero point? Not at all, as it is not the void, but an all-encompassing environment as a superconductive medium. This is the secret of its features that seem unexplainable within the mechanical models of ether (more on that in later chapters).

But unison is not the only synchronization option. There may be other orders of frequency-phase couplings. Suppose the coherent state of the structure of the energy fluctuations of the conductor itself corresponds to the wave structure of the incoming stream. In that case, the stream can pass along this channel with almost no resistance and energy dissipation. Any resistance means getting out of the synchronization region and losing energy transfer efficiency. Harmony creates superconductivity, and harmonic combinations can be different. This is the secret of the phenomenon of superconductivity at high temperatures.

The harmonic ratios can be established under different phase portraits of the incoming wave and the conducting medium. The approach to absolute zero means the establishment of a coherent low-frequency structure through which the external energy flow can pass almost without loss: resistance, as energy dissipation, tends to zero. Under such conditions, even a traditional insulator can become a superconductor. However, such a state can arise not only at a temperature close to zero because the determining factor is not the temperature as such (manifestation of the oscillation frequency) but the ratio of the frequency characteristics of the incoming wave and the conducting vibrational medium. Thus, TEH explains the mystery of superconductivity at different temperatures.

In ordinary conductivity, wave propagation occurs with discrepancies between the system's internal parameters and the parameters of the incoming oscillations, which leads to what we are used to calling resistance (energy dissipation due to ineffective interaction). The insulator does not synchronize with an external source and turns out to be a "breakwater," a current barrier. The superconductor is in a coherent state, and its parameters correspond to the possibility of creating harmony with the external flow. All this explains the different temperature parameters of various materials manifesting superconductivity and that substances can behave like insulators, ordinary conductors, and superconductors depending on the temperature and level of impurities. All is explained by one concept of TEH. There is no need for creating various models for phenomena of one nature and patching holes in them with new virtual entities.

Another effect in superconductivity violates the classical concepts of electrodynamics and is in conflict with SM. This effect was named the Josephson junction after its discoverer Brian Josephson. He suggested that if two superconductors are separated by a thin layer of insulating material, then synchronized electrons will penetrate through it due to the "tunnel effect." If we remain in the corpuscular paradigm, this sounds paradoxical: it turns out that particles easily penetrate through the "walls" of the insulator. This again requires the attraction of magic powers. But for waves, penetration through the medium is a commonplace because waves are not the movement of matter but the transfer of vibrational energy. With certain frequency-phase relationships, the insulator can become a conductor.

Josephson's calculations showed that such a contact works as a voltage-frequency converter, and the wave function expresses the current density. All elements in the superconductor have the same phase, and when the "tunnel effect" occurs, the current flows even without voltage. It was heresy for classical electrodynamics. Imagine that you connected two containers with water, and water flowed between them even in the absence of a difference in their position (zero potential). Josephson and his teachers checked and rechecked the calculations, as they seemed incredible. They showed that the current would depend not on the potential but the phase difference between the two superconductors. In "Josephson generation," with an increase in the phase difference the current grew up to the moment of orthogonality (phase difference of 90°) and then it began to decrease (sinusoidal proportionality).

Josephson junction can be used as a generator and receiver of waves at frequencies unattainable for other methods. It is as if two communicating containers without a difference in position or with a constant difference would generate fluctuations in the flow of water among themselves at a particular frequency. The frequency of the "Josephson generation" is such that it takes the breath away: 100 billion revolutions per second.

There was another blow for the mainstream models of electromagnetism. Josephson junction is not associated with magnetic phenomena, and current induction occurs without the movement of magnets. It is also called kinetic induction: energy is embedded in oscillator carriers' kinetic energy, not in some kind of magnetic field. Classical electromagnetic force disappears, as befits phantoms. What was considered and is still considered by many to be the basis for the emergence of electromagnetic waves (state changes and the interaction of the entities "magnetic field" and "electric field") simply does not happen.

So, reality behaved in a non-standard way from the Standard Model's point of view: current passed through insulators; it is created and grows without voltage; it is in no way connected with magnetism. It depends, as befits the wave process, on the parameters of the oscillations involved in it, on their frequencies and phases. It is no surprise for the wave paradigm, and sheer wonders for the corpuscular.

One of the creators of the BCS theory of superconductivity, John Bardeen wrote in a review of Josephson's work that this supercurrent could not exist. Why? Because it contradicted the dogmas of the corpuscular model of which BCS was a part. His logic was simple: though the Cooper pairs breach many bans they cannot leak through the barrier of an insulator, just because it is an insulator. This is banned. Period. Bardeen was an eminent scientist and a Nobel Prize winner. Josephson was a student. But real science is determined not by ranks but by the search for a model to describe reality. After conclusive experimental evidence, Bardeen was forced to admit his mistake. Was this the collapse of his model? No. The "Particle" team is accustomed to scoring virtual goals and hiding the "Wave" team's real goals under the rug.

It is appropriate to recall a similar story from past centuries. At the dawn of the birth of the wave theory of light, Poisson ridiculed Fresnel's idea, which stated that a bright spot could form in the shadow of an object due to wave diffraction. This simply could not be within the framework of the dominant Newton's corpuscular model of light. Who is Fresnel compared to the great prophet who established the dogma that light is flying particles? How can they go around an object? It was not the model that explained the facts, but the facts had to be consistent with the model at all costs. If they are stubborn and do not want to fit into the "Procrustean bed," then such facts are not worthy of existence; they simply cannot be. Fresnel did not argue with dogma but simply demonstrated this effect at the Paris Academy's next meeting. For the wave theory, it was no miracle. The fact got its rightful place in the collection of confirmations of the model, which had great explanatory and predictive power and did not have to deny reality. But has anything changed over the centuries? Until now, leading theories of physics describe light as the flight of virtual photon particles in the void, but with the correction that they exhibit wave phenomena when passing through the medium.

In 1997, the Josephson effect was discovered in another system with phase coherence: superfluid helium. When the waves slow down and stretch at extremely low temperatures, they begin to coincide and fall into one state of synchronized "superfluid." There is an interesting story on how these vibrations in helium were discovered. After long experiments, the team despaired. Even the most accurate oscilloscopes did not reveal anything. The laboratory leader suggested listening to the signals through the headphones. No one had ever thought of listening to the music of matter, and there were no headphones in the lab. They went to the store, bought headphones, and a "miracle" occurred: they heard a signal with a changing tone (frequency), as the theory predicted. They heard one of the melodies of matter.

But what about the tunnel effect that Josephson referred to when explaining the effect of current going through an insulator at the junction? Tunneling is a natural phenomenon for wave propagation. We use it in all technologies designed to increase the efficiency of energy-information transfer. There is a general term for them: waveguide. It traces its history to the experiments of Joseph Thomson at the end of the 19th century. He proposed a theoretical model of guided wave propagation, which was soon experimentally tested by Oliver Lodge and mathematically described by Lord Rayleigh for acoustic waves. The waveguide is intuitively described as a channel where waves propagate in a specific direction due to limited scattering. The energy loss in the channel is minimal, and without the channel it decreases with distance per the inverse square law. Depending on the characteristics of the channel, it is optimal for waves of a certain frequency.

There are naturally occurring channels. For example, the most famous is the underwater sound channel (SOFAR): a layer of water in the seas and oceans in which ultra-long propagation of sound is possible. The wave repels from the channel's boundaries (different parameters of temperature, pressure, and other characteristics of the medium) and even at low intensity can spread over thousands of kilometers. Whales use this channel to transmit their messages to each other at the optimum frequency for this channel.

We started using different channels before the term "waveguide" was coined. Sound transmission in a rope or an empty pipe has been used in one form or another for centuries. The most famous doctor's instrument, a stethoscope, which has become almost a symbol of the profession, is a classic waveguide. In fact, any transmission line is such a wave channel, directing energy waves. We use this phenomenon in broadcasting, in radars, in fiber-optic communications, and so on.

While modeling the Josephson junction effect, physicists realized that it would be most natural to describe the dynamics of the process in terms of phases and ratios of waves on opposite sides of the insulator. Looking at the essence of the description, they saw that the equation is analogous to the one describing the movement of the Huygens pendulums. At first glance, there is nothing in common between clock pendulums and superconductors. But fundamentally, these are phenomena of one nature. The difference is in scale: a superconducting Josephson contact is the size of a bacterium, and the oscillation frequency is 100 billion times faster than that of a pendulum clock.

Here is another phenomenon that has been and remains the subject of heated debate: "The topic of quantum entanglement is at the heart of the disparity between classical and quantum physics: entanglement is a primary feature of quantum mechanics lacking in classical mechanics ... Einstein and others considered such behavior impossible, as it violated the local realism view of causality (Einstein referring to it as "spooky action at a distance") ... The paradox is that a measurement made on either of the particles apparently collapses the state of the entire entangled system — and does so instantaneously, before any information about the measurement result could have been communicated to the other particle (assuming that information cannot travel faster than light) and hence assured the "proper" outcome of the measurement of the other part of the entangled pair ... This means that the random outcome of the measurement made on one particle seems to have been transmitted to the other, so that it can make the "right choice" when it too is measured" (Wikipedia, "Quantum entanglement").

Of course, Einstein denied the existence of phenomena that did not abide by his prophecy's laws about the absolute and ultimate speed of light. The interaction described by the term "quantum entanglement" goes at speeds far beyond the limit established by the prophet. The author of the above article tries to reconcile this with the Special Theory of Relativity (STR) by stating that information cannot travel faster than light so there should be no communication between particles. But this leads to the paradox: information "seems to have been transmitted."

Within the mainstream paradigm, this contradiction is taken as another quantum miracle that cannot be explained. As with the two-slit experiment that we considered in detail in the previous part of the study, we hear the same old story about the strange behavior of particles that make choices to fool the experimenters. Richard Feynman warned: "Do not keep saying to yourself, if you can possibly avoid it, 'But how can it be like that?' because you will get 'down the drain,' into a blind alley from which nobody has yet escaped" (Feynman, 1965).

But if we get rid of the mystical fog of spooky interactions of particles at a distance in the void and turn to interactions of oscillations in the medium, we see that there is no problem with explaining the entanglement. Let's try to ask ourselves a question: how can it be like that? But not within the framework of the

Wonderland of QM, but within the framework of TEH. Perhaps it will lead to a description without violations of realism, i.e., causation and physical meaning. It was what Einstein was looking for. But the irony is that to return to realism, we have to get rid of his model and quantum model altogether.

It is no accident that the word "entanglement" is used to describe the phenomenon because the interpretation within QM is entangled (confused). The original German word "verschrankung" used by Schrödinger to denote a phenomenon can be translated in another way: binding, coupling. Words are critical: they carry meanings. Trite, but true. Perhaps the change in emphasis from "entanglement" to "coupling" will lead us out of the confusion.

Here it is worth recalling the "sympathy of the pendulums" in Christian Huygens' experiment of 1665 that we looked at in the previous volume. The pendulums of the clocks hanging on the wall synchronized slowly. When he put them on the beam between the chairs, they did it faster. If we put metronomes on a beam based on two rolling aluminum cans, this will happen within seconds due to activity of such an oscillatory system. Still, it is a level of very slow frequencies. Now imagine oscillations with a frequency of sextillions or septillions of cycles per second. It is not easy to write (dozens of zeros after one) and it is even harder to imagine. But it is the reality of interactions that we call fundamental. If we look at them from the sync perspective, we are no longer surprised by the immense speeds of energy transfer.

The experiments show that the interaction happens at speeds that exceed the light speed at least 100,000 times. As the tradition goes, any experiment is interpreted as manifestations of the life of particles and their uncanny behavior. In reality, all experiments are about wave processes. In the simplest case, obtaining entangled quantum states means that a laser of a certain frequency and intensity is directed at a certain nonlinear material.

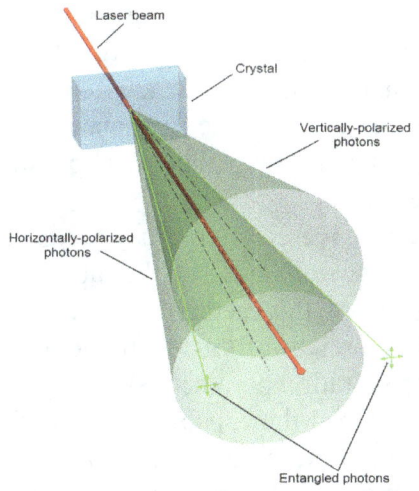

As a result, two polarization cones in a coupled state are produced at the output. This is called "biphotons" or "entangled photons." If not for the usual references to phantom particles, the process's description would be prosaic: the interaction of a coherent light wave (laser) with a nonlinear medium (crystals are usually used) leads to the separation of the wave into two with different polarizations. If we remove the term "entangled photons" and use regular wave formulations "polarized waves" and "synchronized waves," we go out of the quantum fog and get a physically clear picture. The experiment is about creating highly coherent (quasimonochromatic) waves, which can be in a synchronized state of phase coupling (in this case, in antiphase). The presence of a medium ensures the preservation of the frequency-phase relations of the oscillations during their propagation.

The problem of all experiments was one: it was necessary to preserve the initial waves' coherence, despite the influence of various factors, and accurately measure the ratio of the parameters at the beginning and the end of the path. The experiments began at very short distances: in 2007, the record was 1 m. Then they went over to kilometers, and in 2008 they reached a record of 144 km. Coupling remained, and speeds were orders of magnitude higher than light. The mainstream conclusion: the quantum entanglement and, accordingly, the nonlocal nature of reality are once again confirmed.

If we talk about the entanglement of particles in emptiness, it turns out that the nature of reality is non-local, i.e., without normal cause-and-effect relationships. However, in this case, the phenomenon really becomes a "spooky" (read, incomprehensible) action at a distance. If we talk about regular physical synchronization in an active conducting medium, then everything is entirely local, in the sense that there are no gaps in the physical chains of interactions (after all, there is the mediator, not the void), nor in logical chains.

Let's look at the essence of the physics of what is happening in experiments: producing coherent waves with a narrow spectrum; separation of beams into polarized waves; a change in the phase portrait of the oscillations at the input is reflected in the phase portrait of the oscillations at the output; the spread of interaction and state changes depend on the parameters of the sources, medium and receivers; superconducting medium can provide tremendous speeds due to the most efficient synchronization; the frequency range (light waves, electric current, etc.) is not essential, but the ratio of the frequency and phase characteristics of the interacting oscillations is important; thus, it is possible to create conditions for superluminal velocities.

The experimentally discovered enormous speeds of interaction, which can far exceed the speed of light, are explained by the superconducting state, which means the highly coherent state of the medium as a mediator of interaction and the optimal frequency-phase relationships with incoming vibrations. Under such conditions, a change in vibration parameters in one part of a synchronized system can instantly (even faster than the capabilities of modern measurement methods) affect the state in another part. Such a "teleportation" of states is a physically causal phenomenon that does not violate the principle of local realism due to the presence of a continuous energy vibrations transmission chain in the system.

This means that the speed of light is neither the limit for the transfer of energy and information (patterns of energy) nor absolute, as it depends, like any other wave speed, on the parameters of the source, medium and receiver. It also means that the phenomena of instantaneous transfer of the state of a synchronized system from one point to another are not mystical "spooky action at a distance" or "quantum entanglement" of particles in the void, but physical phenomena of wave energy propagation and frequency-phase coupling of oscillations in a medium. This speed is provided by a synchronization mechanism as an effective way of energy transfer.

Let's go back to classical physics when scientists were still investigating physical phenomena in the physical environment and were not shy of making material analogies between different levels of energy interactions.

In 1881, the First International Exposition of Electricity was held in Paris. It presented the most modern achievements, including the Thomas Addison incandescent light bulb, Alexander Bell's telephone, a tram, and an electric car. But no less impressive were the experiments of Carl Bjerknes and his son Vilhelm. They demonstrated the similarity of electromagnetic and hydrodynamic processes. Contemporaries noted: "From a scientific and purely theoretical point of view there is no object in the whole of the Electrical Exhibition at Paris of greater interest than the remarkable collection of apparatus exhibited by Dr. C. A. Bjerknes of Christiania, and intended to show the fundamental phenomena of electricity and magnetism by the analogous ones of hydrodynamics" (Forbes, 1881).

The experiment authors created a design that connected the pumps to the drums (a metal ring covered on both sides with a rubber membrane) immersed in water. As a result of the air supply with a specific frequency, it was possible to create pulsations with certain phase relationships. If the drums contracted and expanded simultaneously (1:1 resonance, 0° phase shift, in-phase vibrations), then they were attracted to each other. If one expanded and the other contracted at that moment (1:1 resonance, 180° phase shift, antiphase vibrations), they repelled. If the phase shift was a quarter of the period (90°, quadrature oscillations), they did not move. It was clear that the oscillators' interaction depended on the frequency-phase coupling (synchronization) and was transmitted through an oscillatory medium.

Bjerknes did another beautiful experiment. He created oscillations of a sphere immersed in water by changing the pressure of the air supplied to it. The author of an article in Nature magazine described it this way: "But in this case it must be noticed that opposite sides of the sphere are in opposite phases. In fact the sphere might be expected to act like a magnet; and so it does. If two oscillating spheres be brought near each other, then, if they are both moving to and from each other at the same time, there is attraction; but if one of them be turned round, so that both spheres move in the same direction in their oscillations, then there is repulsion. If one of these spheres be mounted so as to be free to move about a vertical axis, it is found that when a second oscillating sphere is brought near to it, the one which is free turns round its axis and sets itself so that both spheres in their oscillations are approaching each other or receding simultaneously. Two oscillating spheres, mounted at the extremities of an arm, with freedom to move, behave with respect to another oscillating sphere exactly like a magnet in the neighbourhood of another magnetic pole … Dr. Bjerknes looks upon the water in his trough as being the analogue of Faraday's medium; and he looks upon these attractions and repulsions as being due, not to the action of one body on the other, but to the mutual action of one body and the water in contact with it … The professor shows that if a rectilinear oscillation constitutes magnetism, a circular oscillation must signify an electric current, the axis of oscillation being the direction of the current. According to this view what would be the action of a ring through which a current is passing? If the ring were horizontal the inner parts of the ring would all rise together and all fall together, they would vibrate and produce the same effect as the rectilinear vibrations of a magnet. This is the analogue of the Amperian currents" (Ibid).

The demonstration of the analogy with electromagnetism was not limited to the phenomenon of attraction/repulsion. Bjerknes demonstrated the presence of "lines of force" completely coinciding with classical experiments with a magnet and iron filings. But in this case, the lines were an obvious and direct result of the interaction of the oscillator and the tangible medium. A sphere suspended in water with a brush attached to it drew lines on the glass, reflecting the medium's vibrations. If the drums oscillated in phase, then the lines were similar to those formed by the interaction of the same magnetic poles. In antiphase, the lines were identical to the effects of different magnetic poles. Three pulsating drums gave lines similar to the three magnetic poles:

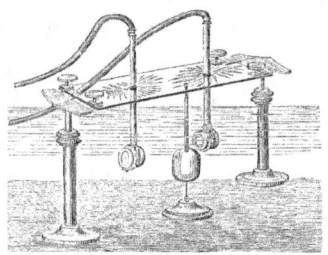

Popular Science Monthly. 1882

They wrote that Professor Bjerknes demonstrated the creation of a field in water. However, the experiment did not demonstrate some "water field" and its lines of force but a phase portrait of waves in water that were completely analogous to the observed phenomena of electromagnetism. However, the reference to forces and fields that "explain" themselves was familiar and convenient. It is for this reason that Bjerknes' experiments, despite initial enthusiastic reviews, became just a curiosity from the exhibition, which was soon forgotten. The mainstream preferred the "good old" fairy tale about the attraction and repulsion of charges in the field.

This tale usually goes like this: once upon a time, people discovered the property of amber rubbed with wool to attract light objects; they called it electron (amber in Greek); then they saw that other objects had this property, and called them electric (similar to amber); when it was discovered that repulsion can occur if other substances rub, people thought that certain spirits lived inside substances, which either were drawn to each other or avoided each other; at first they called them "vitreous" and "resinous" charges after the substances; but many substances had this property, and people decided that there are universal spirits that live in all substances and can move between them; they called them positively and negatively charged particles; when bodies rub against each other the spirits move and the balance is violated so that there is an excess of positive spirits in one part of the body and in the other — negative ones. Children, remember the moral of this story: spirits with the same name repel, and spirits with different names attract.

The story has a sequel. Gradually it was discovered that electrification could occur without direct contact between bodies. At first, common sense seemed to prevail and physicists applied the hydrodynamic analogy that spoke of movement of energy in a medium. In the eighteenth century, Stephen Gray used the idea of "effluvia" as a substance emanating from bodies to describe the process of exchange at a distance. In the same century, Charles du Fay suggested the presence of two fluids, "vitreous" and "resinous," to make a distinction between two types of interaction. Benjamin Franklin reduced the phenomena to one type of fluid. In his model, all substances contain such a fluid, and a different sign means either an excess (positive sign) or a lack of this substance (negative sign).

These were the working hypotheses that sounded quite plausible but lacked an idea about the mechanism behind the interaction. The same question remained: why would things attract or repel? Having two types of fluid inside or some amount of uniform fluid does not settle the issue. The mechanism was not explained, and the old habit of using spirits for the unexplained prevailed. Spirits can do whatever they please: fly between bodies in the void and make them come together or fall apart.

The same tale was taken as a model for magnets, which did not seem to care about any preliminary contact such as friction, and carried out the attraction and repulsion of things through their inherent magical power. It is not some kind of ancient myth. It is the standard modern fairy tale of the Standard Model: "A magnet is a material or object that produces a magnetic field. This magnetic field is invisible but is responsible for the most notable property of a magnet: a force that pulls on other ferromagnetic materials, such as iron, steel, nickel, cobalt, etc. and attracts or repels other magnets" (Wikipedia "Magnet").

It is not a problem if something is invisible. The problem is that the explanation lacks any description of the mechanism of interaction. Thus, it sounds like a description of the magical action of an invisible entity. This is a fairy tale plot. On the other hand, the same article describes the process of magnetizing materials: heating, vibrating and striking the object that is being under an external magnetic influence. This applies to natural magnets that were formed under specific conditions, and those that are produced industrially, when certain configurations of the energy state of the material and external electromagnetic radiation are set. It is about energy interactions, as in the case of electrification from the friction of amber on wool.

Even a preschooler knows that during friction bodies heat up. A high school student knows that it means a transfer of external kinetic energy to internal kinetic (thermal) energy. But it takes a deep dive into real physics to understand that it is all about changes in the amplitude-frequency characteristic and phase portraits of the oscillations of energy that form this particular structure of matter. These changes give rise to various new configurations of frequency-phase coupling, leading to attraction and repulsion, which we observe as a result called magnetism.

Anyone who ever played with magnets knows that they have north and south poles and that opposite poles attract and the same poles repel (analogy of a positive and negative charge). It is again just a name for a manifestation of an interaction direction: north is the one that points towards Earth's north magnetic pole if the magnet is freely suspended and south is, obviously, facing the south pole. As these are just names for directions, they are arbitrary and may even be misleading: the magnet's pole that faces north should be attracted to the south pole of the Earth's magnetic field. So north turns out to be south and vice versa. We should also not take poles literally and think that they are the opposite parts of something. If we break a magnet in two pieces trying to divide the poles, we will find that the pieces "inherit" both poles. This again shows that it is just a name for interaction direction.

The interaction of magnets can manifest itself in attraction or repulsion. They can be in a balanced state or "indifferent" to each other. It depends upon the distance and the properties of the interacting elements. There are many nuances as to the strength and types of magnetism. Ferromagnetic and ferrimagnetic materials are the ones normally thought of as magnets as their interaction is quite strong. Paramagnetic substances are weakly attracted to either pole of a magnet. Diamagnetics are weakly repelled by both poles. There is also superparamagnetism that randomly flips direction under the influence of temperature, superdiamagnetism as the complete absence of magnetic permeability, and metamagnetism as a sudden change in magnetization under small influence of an external magnet.

How does the SM explain all these complicated interactions? "The overall magnetic behavior of a material can vary widely, depending on the structure of the material, particularly on its electron configuration ... When the spins interact with each other in such a way that the spins align spontaneously, the materials are called ferromagnetic (what is often loosely termed as magnetic)" (Ibid).

We have discussed the controversies of the concept of spin within SM in the previous volume. But even if we forget about them, what does "spontaneous alignment" mean? How can particles with the void between them align their spin, be it rotation or any other kind of version of a spin concept? Is it another spooky action at a distance?

But if we use the hydrodynamic analogy and think of energy oscillations in a medium, the magical "spontaneous alignment" becomes a physical phase coupling configuration. As the strength of coupling depends upon the parameters of the interacting elements, distance and the medium state, we can understand why different materials show a wide range of magnetic behavior under various conditions. The crucial point is the presence of a conductive medium where waves propagate and frequency-phase relationships are established between oscillators that can be remote from each other. In this sense, the initial intuition about fluids was much closer to physical meaning than modern theories where interaction becomes the mysterious flight of spirits in the void.

Such a fairy tale is the basis of the most complex models of electromagnetism, where the unexplained magic is hidden behind a fog of terms and formulas describing electric and magnetic fields. For example, there is an experimental fact: two parallel conductors with a current in one direction attract and in opposite directions repel. The force of interaction is directly proportional to the current strength in each of the conductors and inversely proportional to the distance between them (Ampère's force law).

A curious child would ask: why do the same charges repel in Coulomb's law and in Ampère's law electrons with a negative charge are attracted and repelled? On the one hand, there is a similarity (dependence on force and distance), on the other hand, in the second case, attraction and repulsion are determined by the direction of movement of the particles, and not by the type of the charge. Answers of the teachers, textbooks and encyclopedias: Coulomb's law is about motionless point charges in a vacuum, and Ampère's law is about the motion of these charges in the medium. They say that for moving charges additional fields and forces turn on (the magnetic field and the Lorentz force, as the effect of an electromagnetic field on a point particle). As usual, the use of one auxiliary variable entails creating another if ends do not meet.

But the trick does not answer the question. Why do stationary charges obey the rule from the fairy tale about positive and negative heroes, while moving ones simply mock it? If the interaction mechanism in electrostatics is about the attraction and repulsion of positively and negatively charged particles, then what happens to it in electrodynamics? Of course, we can assume that any mechanism works intermittently. However, it is more likely that there is another mechanism that can handle electrostatics and electrodynamics quite well. In addition, the explanation through positive and negative charges is a parody of explanation and does not reveal any mechanism in principle. This is not just an innocent tautology like "water is wet," but a trick that creates an illusion of an explanation that "water is wet because there is wet water in it."

In an attempt to explain, they usually write this: if the conductors have currents of one direction, then the magnetic lines of these conductors, covering both conductors, possessing the property of longitudinal tension and trying to contract, will make the conductors attract; magnetic lines of two conductors with currents of different directions in the space between the conductors are directed to one side and have the property of lateral expansion; therefore, conductors with currents of the opposite direction are repelled from one another:

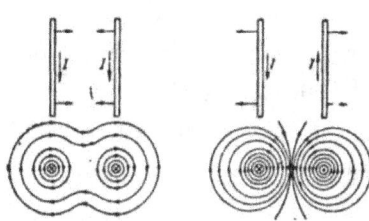

If the particles run in one direction, they create some kind of line of force, spinning together and merging into one compressible configuration. If they run in different directions, these lines collide, expand, and move the bodies apart. If the particles do not run anywhere, then the inherent internal force attracts them to each other or repels them depending on the type of charge. But the same question arises again. Why do same charge type particles (in this case, negatively charged electrons) all of a sudden begin to attract to each other when they move in one direction?

Let's imagine that we look only at the weights of synchronized pendulums in Huygens' experiment and ignore the whole system, including not only the oscillators themselves but also the oscillatory medium between them. We might think that the weights attract and repel on their own for some intrinsic reason. We can attribute these weights a type of charge and say that they repel or attract because they have such a type. It is really very convenient: they go to each other, so different types are attracted; they go away from each other, so the same types repel; if they begin to interact not according to the rule of the charge type, we'll come up with another rule.

We can come up with special fields, forces and virtual carriers of these forces to explain how these weights interact at a distance. We can create complex formulas and diagrams. They will be very convenient phenomenological descriptions because they will reflect the observed. Still, they will not bring us a step closer to understanding the essence of the ongoing process. Since, according to the logic of such modeling, phantoms begin to multiply, they will lead us into the labyrinth of the Minotaur without the saving Ariadne's thread. We will go around tautological circles and run into deadlocks, hiding a simple fact: we're lost.

But it is worth remembering what Huygens did when he discovered the unexpected "sympathy" of the pendulums, which synchronized as if by a wave of a magic wand. He remained a physicist. He did not refer to spooky action at a distance, spirits, or even special fields, but began to study the entire system to understand the mechanism. As a result, he discovered that the secret is simple: there is an oscillatory medium that links the oscillators and allows the process of energy exchange to occur. This mechanism, called synchronization, is Ariadne's thread in the labyrinth of fundamental interactions, although its tip began at such a mundane level and was discovered almost by accident. Now all that remains is to pick it up and not let it slip out of our hands.

If the audience at the exhibition of 1881 had remembered Huygens' pendulums of 1665 and drawn an analogy with drums in water, they would have understood that the point is not in some special "water field" and certainly not in the positive or negative charge of the drums, but in a universal mechanism of interaction. From here it would be a stone's throw to providing a physical basis for the analogy with electromagnetic phenomena that arose naturally due to the same manifestations of interactions. From there, the analogy with other fundamental interactions is not far away (more on this in the following chapters).

But it was already customary to attribute the effect of attraction/repulsion to the charge type difference/coincidence. The mainstream was satisfied with the illusion of the answer. So, when Bjerknes demonstrated that the real mechanism might not be related to the type of charge, everyone was puzzled. Moreover, at first glance the hydrodynamic analogy looked like "anti-Coulomb": in-phase oscillators attracted and antiphase repelled, while the same charges should repel and vice versa. The confusion of the real category of the oscillation phase with the virtual category "charge type" gave rise to bewilderment: maybe the phenomena just look similar but are not analogous. At the same time, there was a complete analogy with Ampère's law. It looked like a dead-end. In fact, it was. The only reason is the concept of the charge type.

Let's remove this concept and put the concept of the oscillation phase into both laws. Then attraction will mean not the difference in charge type, but phase coincidence, and repulsion — not a coincidence of charge type, but phase difference. The contradiction disappears by itself. It's all about the waves in the energy medium and their frequency-phase interaction. Moreover, the wave nature of the observed phenomena explains the dependence of the strength of interaction on distance both for moving and static charges. It can be intuitively illustrated by a simplified geometrical representation of point-source radiation in three-dimensional space:

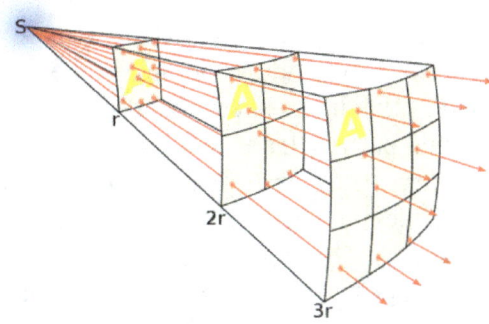

The illustration shows how radiation passes through a surface and its area increases in proportion to the square of the distance from the source, and the intensity of radiation passing through the same area decreases inversely to the square. Of course, this law again should be considered an approximation and should have necessary corrections for the nonlinearity of processes and assumptions about the point nature of the source.

If we think of interactions as the exchange of virtual particles in the void, we are puzzled about the reason for the dependence upon the distance. What can reduce the power of angels' wings? But if we get down to the material world and talk about physics and not miracles of the quantum Wonderland, we understand that wave propagation in a tangible medium depends on the source and the medium's parameters. The more absorption and scattering in a particular medium, the less the intensity. The law will not work in its linear proportionality if there is a prevention of energy dissipation during wave propagation in a medium (for example, in a waveguide). Here the hydrodynamic analogy works again: in a water channel, the flow's intensity is not defined by the simple inverse square law.

Hypothesis:

The attraction and repulsion of conductors with the same or opposite direction of electric current (Ampère's force law) is the result of in-phase and anti-phase coupling of longitudinal and transverse electromagnetic waves. The same phase relationship is behind the attraction and repulsion of standing waves (static electrical charges in Coulomb's law). There is no difference between electrostatics and electrodynamics as the mechanism of interaction of waves in the medium is the same and depends on the ratio of frequencies and phase coupling configurations. For both cases the force of interaction is directly proportional to the current strength and inversely proportional to the distance, indicating the wave nature of electromagnetism.

This hypothesis gets rid of phantoms of SM and explains the observed phenomena with a physical mechanism in a material world. This mechanism works for all electromagnetic phenomena as well as for hydrodynamics, acoustics and other wave interactions. We can use hydrodynamic analogy or turn to musical analogy as we did in the previous volume. They are interchangeable as they speak about the same physical mechanism. As we will see in the further chapters, they work in tandem for the rest of the fundamental interactions (gravitational and nuclear).

The concept of charge is neutral in the sense that it can be used both in the corpuscular and wave models of the world. It can be interpreted both as the amount of energy in a discrete particle and as the amplitude-frequency characteristic of oscillations. What makes the difference it that thinking about charge as the energy of a discrete fundamental particle leads to internal contradictions of the model and its contradictions with reality. Thinking about it as a quantized measure of continuous waves and their amplitude-frequency parameters gets us out of the dead-ends of the maze.

The particle paradigm postulates rules of the "ball game" where point charges produce fields and interaction of these charges with their product results in an electromagnetic force as the fundamental interaction. It is mediated by photons flying in the void between charged particles that make the field which does not exist in the void but should be everywhere else to make things work. This conceptual mess is called Quantum Electrodynamics (QED). It is impossible to make sense of this mutual procreation of fields in the void with no fields. But no one said that it should make sense. Moreover, its founding fathers stressed that we should not look for any sense where there is none, and trying to ask questions will lead us down the drain.

Generations of theoretical physicists listened to this advice and continued to produce non-physical nonsense resulting in a Quantum Field Theory (QFT) and its general framework called the Standard Model (SM) that tries to combine all fundamental interactions. Here is a short description of how things work as per these "physical" theories: pairs of virtual particles and antiparticles are constantly born and annihilated in the void, creating quantum fluctuations of the fields associated with these particles; these virtual particles with opposite charges and different quantum numbers are responsible for all fundamental interactions at a distance in the void. It would seem that the word "fluctuation" which has connotations of the oscillatory process should carry physical meaning. But as the postulated mechanism of interaction is still the exchange of virtual entities flying in the void, we lose this glimpse of sense and it vaporizes into this great nothing.

Here is a typical example of an explanation of a real physical phenomenon of interaction. When two uncharged conductors are placed in a vacuum close to each other (distance of nanometers), they attract or repel each other depending upon the position and the strength of interaction depends on the distance. This is called Casimir effect. In essence, it is a typical interaction with usual manifestations. The mystery was the lack of connection with the type of charge, since the conductors are not charged. The QED, QFT and SM rush for help with explanation: "Using the quantum electrodynamic vacuum, it is seen that the plates do affect the virtual photons which constitute the field, and generate a net force — either an attraction or a repulsion depending on the specific arrangement of the two plates. Although the Casimir effect can be expressed in terms of virtual particles interacting with the objects, it is best described and more easily calculated in terms of the zero-point energy of a quantized field in the intervening space between the objects" (Wikipedia, "Casimir effect").

Isn't it clear? We can call for virtual spirits at any time to help us explain any virtual field in the void that generates real phenomena observed in the material world. Spirits interacting with reality is always a good explanation, but we can also use zero points of emptiness to calculate everything easily using mathematical tricks. The founding fathers who invented the tricks understood that it was a shell game (fraud) and warned: "Having to resort to such hocus-pocus has prevented us from proving the theory of quantum electrodynamics is mathematically self-consistent" (Feynman, 1985). But who cares about the mathematical inconsistency of the world of spirits?

This is how spirits work with reality in the case of the Casimir effect according to SM: the pressure of virtual photons from within two surfaces where their birth is suppressed is less than the pressure from outside where their birth is unlimited; this "negative pressure" of spirits in the emptiness that has become even more empty causes the plates to attract. This is not a joke. It is the description and explanation within the main "physical" theory of the 20th century that speaks of ghosts that are born without any limits on the outside but are aborted on the inside. Alice would wake up from such a dream in a cold sweat and jump out of the rabbit hole. But for many theorists, being in a quantum Wonderland is a normal state of mind.

If we wake up to reality, we will see that the Casimir effect depends on the interaction of waves in a medium, and not on the pushing of plates by spirits in the void. At certain frequency-phase relationships, interference can be constructive, and at others destructive. Electromagnetic waves between the plates can be amplified or suppressed. This is what causes the observed motion of the plates, which we interpret as mutual attraction or repulsion, when in fact it is a process that occurs in a system that includes the environment around the plates. Technical vacuum produced in experiments is a medium with no gases, but it does not mean that it is an empty space with no physical quantities. We can again take the hydrodynamic analogy to illustrate the mechanism of the Casimir effect. If two ships get side by side, the interference of waves in the space between them can suppress the waves. The calm sea between the ships creates less pressure than the waves from the ships' outer sides. As a result, the ships "attract." Do we need virtual ghosts pushing ships together? No. Even sailors of the old times did not call for spirits to explain such an event. Why? They saw water and waves in it with their own eyes. With electromagnetic effects, sometimes you do not see the medium and waves, but only their interaction manifestations. Should we use the analogy for a physically comprehensible explanation or should we call the spirits of the void?

We can even combine hydrodynamic and acoustic analogy. If we place two plates in a tank filled with water and convert sounds into water waves, the plates will be attracted or repelled depending on the wave parameters. It is a version of the Bjerknes experiment, but instead of air pumps creating oscillations of the drums or spheres, a sound source creates air vibrations interacting with water.

If we model hydrodynamic and acoustic effects as some special fields that somehow interact, we either get into a magical thinking mode (fields or forces do something in their unknowable way) or into a tautological circle (fields interact this way because they are fields that interact). Both versions work separately and can co-exist in the same model but they do not have any scientific content as the physical mechanism of interaction remains unclear. However, when it comes to hydrodynamic or acoustic phenomena, we are used to modeling them as waves because the medium of their propagation is known. But we have to remember that for a long time aerodynamic and acoustic effects were also a mystery as air is not visible and available for direct inspection like water is. Only progress in indirect methods of studying the composition of the air led to an understanding that it is not an emptiness where things happen in a magical way but a tangible medium where oscillations and waves interact. Moreover, knowledge of the regularities of acoustics due to the development of musical scales and sound production technologies led to ideas about the underlying frequency-phase coupling mechanism. This is a base for a musical analogy spreading to all acoustics and from there to hydrodynamics.

With the fundamental interactions (electromagnetic, gravitational, nuclear) the medium is also not available for direct study at the current level of our technologies. But this does not mean we should get back to magical thinking or swirl in tautological circles about fields. We do not use the concept of a field as some kind of mysterious force acting on things when it comes to hydrodynamics, aerodynamics, or acoustics. We talk about waves in the medium. But as far as fundamental interactions are concerned, we are still in the grip of mystical awe. Moreover, we model them as some magical deeds of phantom entities. We do not think of them as waves in a medium but flights of virtual particles in the void or curves of space-time fabric. This is just to hide the fact that we cannot explain the actual physical mechanism of interaction. TEH offers the way out: spreading musical analogy to all interactions on all levels of energy without dividing them into fundamental and non-fundamental ones.

Let's consider some more examples that show how various levels interact and demonstrate the intersection of what we consider typical wave phenomena in a medium with what the mainstream theories model as virtual entities acting in the void.

Can acoustic waves counter-balance gravity? If we think of gravity as the result of curves in space-time fabric (General Theory of Relativity) or bodies exchanging virtual gravitons (Standard Model), we are sure to say that it is impossible. Reality says that it is possible. There is an acoustic levitation technology. It is real and has a long history. But the practical use does not mean that the phenomenon has been explained. The article in Wikipedia is concise and does not dwell on process physics at all. It merely states: "Acoustic levitation is a method for suspending matter in air against gravity using acoustic radiation pressure from high intensity sound waves" (Wikipedia, "Acoustic levitation"). It lists experiments and pragmatic aspects of the application. They include contactless manipulation of molten or corrosive materials, micro-assembly, and 3D printing.

What is the principle of the devices? We need a radiator, a reflector, and a matching frequency of sound so that a standing wave arises that compensates for gravity. Such acoustic pressure is already capable of lifting objects of several kilograms. Very accurate and focused synchronization of acoustic waves is required for the power to be sufficient. This is similar to a laser, as a device where the pump energy (light, electric, thermal, chemical, etc.) is converted into a highly coherent monochromatic energy wave. The formation of such a coherent structure of acoustic waves is understandable and even banal from the point of view of wave theory.

A standing wave is one of the manifestations of interference when coherent waves propagating in opposite directions create an alternation of maximums (antinodes) and minimums (nodes) of amplitude that are stable in a spatial arrangement. Oscillations of a string resonating with a musical instrument's body or air vibrations in an organ pipe are classic examples of such a standing wave. A vivid natural example is the "Schumann resonance": the formation of standing electromagnetic waves of low frequencies between the Earth's surface and the ionosphere. It is a massive resonator in which coherent waves create a stable resonance and interference pattern. A similar picture appears in the acoustic levitator, and the object begins to shift towards the nodes. Where the pressure of the acoustic wave balances gravity, the object stops. With pressure dropping, it begins to move. Thus, it can be manipulated by changing the parameters of acoustic waves.

If we apply the basic postulates of SM, it turns out that real acoustic waves interact with virtual gravitons to compensate the force of gravity. If we use GTR's postulates, it turns out that the curve of the space-time fabric adjusts to acoustic waves in such a way as to compensate for gravity and lift the object. It sounds almost comical. The article in Wikipedia does not mention any theoretical work except for acoustic theory of Lord Rayleigh, which did not say anything about acoustic levitation, and some mathematical descriptions of the effect. In other words, there is no model for a phenomenon that indicates the intersection of wave interactions of different energy levels. For mainstream models, gravity-acoustics is just another "curios thing from the exhibition," and they simply ignore it.

By the way, let's return to the Paris exhibition in 1881. There was another curious thing there in addition to the Bjerknes experiment. Many people know that Alexander Bell exhibited his phone, which was already technologically mastered, patented and sold. But few people know that a year earlier, Bell created another device, which he considered his major invention: a photophone. It was a wireless device where information was transmitted by a light beam modulated by the sounds of the voice. The action was based on the property of selenium to change the conductivity in the rays of light. The transmitter consisted of a thin mirror vibrating under the influence of a sound wave. The medium of energy-information distribution was air. The receiver had crystalline selenium cells, which changed their conductivity depending on the light-shade pattern in which the information was encoded. It is a complete analogy to a conventional telephone. The difference was only in the carrier and the medium of wave propagation. Bell's photophone was the prototype of all modern wireless communication technologies, but he was ahead of time: there were severe restrictions on distance, signal visibility and efficiency. We can also say that it was also a prototype of all fiber-optic communications since they use a modulated light signal.

The photophone became another forgotten curious thing from the exhibition, and Bell became the unrecognized founder of optical acoustics or acoustical optics. No matter which side you look at, it was about the interaction of light and sound, i.e., physically about the transformations of different frequency levels of energy oscillations into each other and technologically about the transfer of information, encoded in the amplitude-frequency-phase pattern of these oscillations. Light can affect the propagation of acoustic waves, and sound can influence light phenomena. With the help of sound, we can control the scattering and refraction of light, the laser intensity, the beam's path in space, polarization, spectral composition and structure of optical rays. Powerful light waves can amplify sounds and even generate them. Sound can, in turn, generate powerful light. How is this possible? No problem if we talk about the interaction and synchronization of vibrations at different energy levels. It is possible to describe processes physically and with normal cause-effect relationships.

A vivid (in literal and figurative sense) example of the interaction of oscillations at different frequency levels, which we are used to separating across various "fields," is the phenomenon of sonoluminescence. The name speaks for itself: a glow generated by sound. This phenomenon appears on the list of unresolved problems of modern physics, as an example of how existing theories are unable to explain observed phenomena or experimental results.

Wikipedia formulates a description in this way: "Sonoluminescence is the emission of short bursts of light from imploding bubbles in a liquid when excited by sound" (Wikipedia, "Sonoluminescence"). It is the appearance of a flash of light during the collapse of cavitation bubbles generated in a liquid by a powerful ultrasonic wave. A typical experiment of observing sonoluminescence is as follows: a resonator is placed in a water container and a standing spherical ultrasonic wave is created in it. With sufficient ultrasound power, a bright point source of bluish light appears — the sound turns into light.

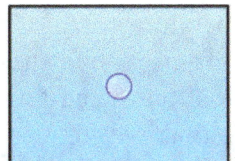 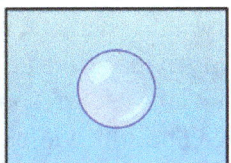 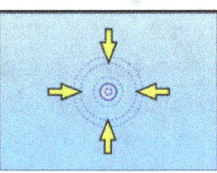 

This is an empirical fact. The only thing left is to explain. Let's look into details of the phenomenon. What is cavitation and cavitation bubbles? We can observe these bubbles every day when we boil water for tea or coffee. When water is heated, the pressure of its saturated steam rises. At the boiling point, the pressure becomes equal to atmospheric, and bubbles appear in the water. This means that the pressure at different points of the liquid has become equal to saturated steam pressure. The liquid in this part evaporates, and a vapor bubble forms.

But what does the sound have to do with it? Such bubbles can form when exposed to sound waves at specific frequencies. But furthermore, the usual bubbles turn into "mini-suns": they glow as bright as the Sun literally. This phenomenon was discovered back in the 1930s by physicists from the University of Cologne. Still, since it was a single and dull flash, the phenomenon did not attract too much attention. Only half a century later, the phenomenon manifested itself brighter: it became possible to create a bright, continuous and stable light, to study the glow with good resolution capabilities of the equipment.

It turned out that a standing ultrasonic wave in the rarefaction phase creates low pressure in the water, leading to local rupture and bubble formation. Over a quarter of the wave period, the bubble grows. If the standing wave is spherical, then the bubble is spherical. In the next phase of wave compression, the bubble collapses, and at the end of the phase, a short and bright flash of light appears from it. At a certain frequency of the standing acoustic wave, stable synchronization is created, and the bubble appears, flashes and disappears with a constant frequency. The bubble stops in the region of the node of the standing wave (minimum amplitude). The resonance frequency depends on the shape, size of the container and the medium characteristics. For example, if inert gases are added to water, the effect is enhanced. The frequency of such outbreaks (births and deaths of the bubble) is very high and amounts to millions of hertz. To fix the flash time, even modern equipment's resolution power is not enough, and only an upper limit of 12 picoseconds has been set (a picosecond is equal to a trillionth of a second). As a result of discrete flashes with an enormous frequency, we see constant sonoluminescent light with a fairly smooth spectrum. The emission spectrum is related to the frequency of oscillations, and the frequency, as an energy category, is directly related to temperature.

Experimenters noticed that with sonoluminescence, the intensity increases strongly in the violet side of the range (the limit of the high-frequency spectrum for our perception). This became a revelation and, at the same time, a stumbling block in trying to explain the phenomenon. The spectrum indicated that temperatures reach a range of 2300 to 5100 Kelvin (for comparison, the Sun's surface temperature is about 5780 K). "A mini-sun in a glass of water" is not a metaphor but reality. Furthermore, color temperature calculations led to astounding numbers of radiation intensity. At a water temperature of 22° C, the spectrum of sonoluminescence corresponds to the spectrum of a black body heated to 25,000 K, and when water is cooled to 10°, it corresponds to a spectrum of a black body in the range above 50,000 K. The ultraviolet and above frequencies are absorbed by water; therefore, the observed spectrum may indicate temperatures over a million Kelvin, i.e., the level of thermonuclear reactions within stars (Moran et al., 1995). Some authors wrote about evidence of nuclear radiation during acoustic cavitation (Taleyarkhan et al., 2002). Other researchers said that a plasma core appears in the bubble with an ionization and excitation level of 18 electron volts, indicating temperatures of at least 20,000 K (Flannigan, Suslick, 2005).

It began to resemble a source of infinite energy. When comparing the low power of the acoustic wave that causes the energy of the thermonuclear intensity level, it turned out that the energy concentration in a flash reaches about a trillion times. Such a staggering picture led to ideas about controlled thermonuclear fusion "in the kitchen." Of course, we cannot directly measure the temperature inside these bubbles of a fraction of a millimeter. But we calculate the temperature of the Sun and other stars from the emitted spectrum. We can do the same for mini-suns. Regardless of the accuracy of color temperature calculations, the fact remains: the frequency of sonoluminescence oscillations is orders of magnitude higher than the frequency of the sound wave that causes it.

The phenomenon was full of paradoxes from the point of view of standard theories and got onto the list of unresolved problems. Wikipedia writes simply: "The mechanism of the phenomenon of sonoluminescence is unknown" (Wikipedia, "Sonoluminescence"). Among the main hypotheses, the following are listed: bremsstrahlung radiation emitted by a charged particle when it is scattered; corona discharge in a stressed field; radiation from particle collisions; proton tunneling effect (overcoming by a microparticle of a potential barrier); non-classical light (special radiation of a quantum field); electrodynamic or fractoluminescent jets and so on.

All these hypotheses using the usual concepts of fields and particles required a variety of magical transformations. Still, even such standard tricks do not help in answering the most important questions. How does a sound wave cause frequencies with a completely different energy level? Where does such a fantastic conversion scale come from, and what is its mechanism?

Attempts to explain use familiar concepts of quantum fields and virtual particles. For example, the hypothesis developed by Claudia Eberlein: everything happens due to coordinated oscillations of a quantum vacuum filled with randomly interacting virtual particles; the bubble boundary moves so fast during compression that it turns virtual vacuum photons into real photons of light flashes

(Eberlein, 1996). Some virtual particles are in chaos, but the compression of the bubble somehow organizes them so that angelic spirits in the void turn into real light in water. What is the mechanism "under the bonnet"? The answer: the field of quantum miracles.

The mysteries of sonoluminescence include the amazing stability of the phenomenon: the bubbles' steady position and the strict periodicity of flashes. Moreover, the flash frequency is more stable than the frequency of the sound generator. Here it is appropriate to recall the younger Bjerknes, who empirically revealed the so-called "primary and secondary forces of Bjerknes." Continuing his father's work, Vilhelm studied the behavior of bubbles in a liquid and found that when exposed to external sound vibrations, specific interactions are established between the external waves and the bubbles themselves (Bjerknes, 1906).

Primary force (out of habit, the phenomenon was immediately called "force") showed standard patterns: attraction/repulsion depending on the phase shift and inversely proportional to the distance. But between the bubbles themselves, phase relations were also established, which prevented the collision and collapse. Bubbles with an oscillation frequency higher than the carrier acoustic wave moved upward along the pressure gradient and downward with a low gradient. In a standing wave, small bubbles tended to nodes and large to antinodes. In general, complex but steady dynamics were observed, similar to many synchronized systems. The analogy with space macrosystems and atomic microsystems is obvious: synchronization can form stable configurations with an attraction and repulsion equilibrium.

Bjerknes' experiments with bubbles clearly show that we can perceive the phenomenon as a separate particle, and interaction as the ratio of particles, while it is part of a single system, and the interaction has complex nonlinear dynamics of all energy parameters of the system. A bubble in water seems to be a separate object, although this is a formation within a single environment. It appears that the balls are attracted and repelled, although these are just phases of oscillatory dynamics within the general phase portrait of the entire system. Now imagine that we do not see the medium, but see only the bubbles. We may create the Standard Model of particles' interaction, come up with magical forces and virtual particles exchanged between the bubbles to attract and repel. The appearance and disappearance of bubbles we describe as a supernatural birth from nowhere and annihilation into nowhere. We persistently model all possible combinations and dynamics as interactions of virtual particles in the void and draw diagrams of such combinations. In a sense, this is a convenient description, but we do not get anywhere close to the physical meaning of the process. Moreover, we get further and further from it into the blind alley of phantoms in the void. This is what happened in the 20th century and continues to these days.

Carl Bjerknes dedicated 20 years to building a theoretical model of his experimental results. But the problem remained: it did not reveal the mechanism, but only mathematically described the observed manifestations. Experiments have demonstrated patterns that indicate the universal nature of energy interactions. But to draw conclusions about real similarity, it is not enough to describe the observed.

The ultimate goal is to explain the mechanism. As a result, the theory was not refuted but simply forgotten. Moreover, it clearly said that no forces are acting at a distance in emptiness, and all phenomena are manifestations of the interaction of energy processes in a real medium. A few decades later, such a simple idea became blasphemous to the new church of mainstream theoretical physics. That is why Bjerknes experiments with far-reaching conclusions about the general laws of interactions disappeared in the mainstream swamp. Attracting, repelling and creating stable patterns bubbles gurgled somewhere on the periphery and dissolved. As well as the glow of the bubbles, discovered later. New data at the end of the century on the cosmic scale of transformations in the glass of water came as a shock, but it was also absorbed. The topic is on the margins of science. Why? The phenomenon cannot be explained using standard theories.

When considering the phenomenon of sonoluminescence from the point of view of the laws of wave processes and synchronization of oscillations, it becomes surprising: why do we need some exotic "non-classical fields" and quantum miracles when trying to explain? Only out of the habit of inventing phantoms and producing entities. But if we get rid of this bad habit, we can combine different phenomena into one concept. Let us try to formulate an explanation for such a mysterious phenomenon for modern theoretical physics as sonoluminescence. It is highly probable that the mystery will evaporate as water in a cavitation bubble.

The huge scale of the conversion of a sound wave into high-frequency oscillations of the light range in the effect of sonoluminescence is explained by phase-frequency coupling. An external sound wave of a specific frequency during synchronization causes a response of internal oscillations of the medium (in this case, water) and generates a resonant effect, which can have a vast difference in scales between the source of oscillations and the resulting frequencies. Sound does not create light; it causes it. Since these are wave phenomena, they "speak the same language." Energy is not generated from nowhere but is contained within the environment.

A "golden key" to this door is synchronization. The stability of the phenomenon is explained by the fact that a dynamic system of synchronized oscillations, in which several levels participate, including the internal oscillations of the propagation medium itself, is established. Such a system can form stable structures, convert energy from one frequency level to another, and develop tremendous speeds, frequencies, and power. The interaction of oscillators in an environment, in which certain phase-frequency relationships are established, allows them to transfer and convert energy creating wave patterns. An analogy with electromagnetic, gravitational and other interactions suggests itself. Such a simple but far-reaching conclusion is far from the mainstream paradigms, and they habitually produce virtual entities and describe their magical transformations.

Let us return once more to the initial Bjerknes experiment as a clear demonstration of the general laws of such a process. Of course, it had a minimal range of parameters: the frequency ratios were the simplest (1:1), and the phase relations were reduced to phase, antiphase and a quarter of the period. If we represent a phase portrait of the interaction of oscillators with the same oscillation frequency and with a phase ratio of 0, 90 and 180 degrees in the form of a Lissajous curve, then the picture will be as follows: in the first and third case (in-phase and antiphase), the phase point moves in a straight line back and forth; in the second — it draws a limit cycle (circle).

We can say that the observed phenomenon of attraction, repulsion, or the equilibrium state of oscillators in a medium is a phase portrait of their synchronized interaction. When the phases are perpendicular, the oscillators are in an equilibrium of the limit cycle, and the wave picture of the transfer of this energy in the medium establishes a push-pull mutual balance. When the phases coincide or are opposite, the motion of the oscillators creates energy transfer in one of the directions, the state of the medium changes, and the oscillators themselves begin to attract or repel precisely due to a systemic phenomenon in which all the elements participate (the objects themselves and the wave propagation medium).

The stability of such a system is determined by the synchronization band and the state of the medium. When we talk about a phase portrait in the form of a straight line, a circle, a lemniscate and other variants of a stable and coherent state, we must not forget that only ideal relationships can "draw" ideal figures. In reality, there is a band (Arnold tongue) within which interaction can be established. Recall that the main parameters determining the bandwidth are the frequency-phase relations and the amplitude of the impact.

In such a model, it becomes understandable why such seemingly different phenomena exhibit universal laws. They depend not only on the parameters of the interacting elements themselves but also on the medium's parameters, which naturally affects the amplitude of the action and the coupling force. It works for drums in water, for bubbles, for electricity, for light, for gravity and other types of interactions.

CHAPTER 4

THE PUZZLE OF GRAVITATION

The history of science shows that the progress of science has constantly been hampered by the tyrannical influence of certain conceptions that finally came to be considered as dogma. For this reason, it is proper to submit periodically to a very searching examination, principles that we have come to assume without any more discussion.

Louis de Broglie

We have dwelled upon the enigma of gravitational interactions in the previous part of the study. We have also started to reveal the secrets of nature that underly the phenomena of energy interactions from microcosmic to macrocosmic levels and showed that they are all based on the common mechanism called synchronization or frequency-phase coupling of oscillations. The music of the atomic and celestial spheres is based on the same laws of physical harmony. In this volume, we continue the research and develop the Theory of Energy Harmony (TEH) that aims to solve the puzzle of fundamental interactions. It unifies all known energy processes in one model. This is the ultimate task of theoretical physics. Unfortunately, the mainstream has substituted this task for another one: to unify the existing models of various interactions. This approach stems from a belief that these models are correct and all that remains is to somehow reconcile their contradictions with each other. Internal contradictions of the models and their contradiction with reality and physical sense do not bother the mainstream. To be precise, those who are bothered are ostracized and expelled from the "scientific" community which has become a religious corporation based on the beliefs in virtual spirits responsible for material world phenomena and fiercely defends its dogmas. This is actually the main problem of any attempt to unify these models in one theory of everything. They speak of different intangible entities. Thus, the gods of these religions cannot be reconciled within one church.

This may sound surprising and even offensive to those who sincerely believe "holy scriptures" of the mainstream theories. But there is no offense in taking a religion for what it is: a worldview that is based on a belief in non-material entities. Whether or not to share this view is a matter of personal choice. From a psychological point of view, this is just a strategy to cope with the fear of the unknown. However, belief in the immaterial should not be part of physical science. These are incompatible models of the world. Any attempt to introduce an element of faith into science generates cognitive dissonance among believers themselves. That is why they are offended when the real state of affairs in their cherished models is pointed out. After all, they call them physical theories, and themselves physicists. Obviously, this "ear-shutting" tactic will not help to get rid of dissonance in the long run. Meanwhile, it is a great hindrance to the development of science, so we have to continue to voice the truth in the hope of being heard.

In this chapter, we will speak of how the two main models of theoretical physics, the General Theory of Relativity (GTR) and the Standard Model of particle physics (SM), look at gravitational interactions. Even at the risk of repetition, we will show their deficiencies and contradictions just to pave the wave to a new worldview based on a truly scientific approach that is not afraid of the unknown and does not cover the holes in the explanatory base by inventing immaterial entities.

We will start with Einstein's theory. Here is how mainstream worships the prophet of its religion: "General relativity, also known as the general theory of relativity and Einstein's theory of gravity, is … the current description of gravitation in modern physics … , providing a unified description of gravity as a geometric property of space and time or four-dimensional spacetime. In particular, the curvature of spacetime is directly related to the energy and momentum of whatever matter and radiation are present. The relation is specified by the Einstein field equations, a system of second order partial differential equations. Newton's law of universal gravitation, which describes classical gravity, can be seen as a prediction of general relativity for the almost flat spacetime geometry around stationary mass distributions. Some predictions of general relativity, however, are beyond Newton's law of universal gravitation in classical physics. These predictions concern the passage of time, the geometry of space, the motion of bodies in free fall, and the propagation of light, and include gravitational time dilation, gravitational lensing, the gravitational redshift of light, the Shapiro time delay and singularities/black holes. So far, all tests of general relativity have been shown to be in agreement with the theory. The time-dependent solutions of general relativity enable us to talk about the history of the universe and have provided the modern framework for cosmology, thus leading to the discovery of the Big Bang and cosmic microwave background radiation. Despite the introduction of a number of alternative theories, general relativity continues to be the simplest theory consistent with experimental data. Reconciliation of general relativity with the laws of quantum physics remains a problem, however, as there is a lack of a self-consistent theory of quantum gravity. It is not yet known how gravity can be unified with the three non-gravitational forces: strong, weak and electromagnetic.

Einstein's theory has astrophysical implications, including the prediction of black holes — regions of space in which space and time are distorted in such a way that nothing, not even light, can escape from them ... Other predictions include the existence of gravitational waves, which have been observed directly by the physics collaboration LIGO and other observatories. In addition, general relativity has provided the base of cosmological models of an expanding universe. Widely acknowledged as a theory of extraordinary beauty, general relativity has often been described as the most beautiful of all existing physical theories" (Wikipedia "General Relativity").

You need to take a breath after such breathtaking praise of an extraordinarily beautiful and simple model that not only describes and explains gravity but also makes predictions about the entire Universe, which have been confirmed more than once by experiments and observations. The only cloud in the sky is that it cannot be reconciled with the models of other fundamental interactions. But it does not seem to be a problem because alternative models are not self-consistent. Why reconcile the beautiful with the ugly? The small trouble remains: all fundamental interactions demonstrate the same patterns hinting at the same mechanism but models speak of different ones. Of course, the author of the article does not mention the main trouble: all these theories, including the beautiful and simple TR, employ transcendent immaterial entities to explain physical reality. But as we have mentioned earlier, this is no trouble at all for the mainstream that is used to metaphysical and religious content of its cherished dogmas.

But let's look at the postulated great insights of Einstein and their claimed experimental confirmations. We have revealed the absurdity and incompatibility with the reality of some of his ideas in the previous part of the study. We have also shown that experiments either did not have anything to do with confirming these ideas or even directly contradicted them. Here we will limit ourselves to the ones related to gravitational interactions. We will find the same trend: what propaganda calls a beautiful theory confirmed experimentally is a conceptual mess detached from reality.

We will start at the beginning: the main idea of GTR is that gravity is not an interaction at all but a result of the curvature of space-time fabric leading to the observed phenomena that we take for the interaction results. According to Einstein, it is our illusion that things interact. What actually happens is that everything moves along the curves of this universal entity that is bent like an old sofa. It is really very simple. Even a child can visualize. But to help us, they draw pictures like this in textbooks:

In the GTR framework, the trajectory of objects is explained by the "geodesic paths" of the geometry of the space-time fabric. However, it is an experimental fact that gravity depends on the mass of objects. So, GTR states that objects affect the geometry of the fabric by their mass. Thus, the bodies sit in the curvatures of the sofa, but they also bend the sofa. It is logical: if the sofa is curved, then you will sit in this curvature. But the original premise was that your sitting configuration is determined by curvature. Who curved it? We return in a circle: a sitting body. But why is it sitting in such a crooked way? Again, in a circle: because the sofa is curved. It is a logical error: a cycle of logic in which the initial premise (the presence of curvatures) explains itself. As a result, GTR supporters can write within the same article or even a paragraph that gravity is created by curvature and that curvature is created by the gravitational effect. They do not even notice this circle of logic. But as we have noted in other cases, such circles are inevitable when the mechanism is not disclosed. After all, even if we accept the main GTR hypothesis that the curving space-time produces gravitational effects, what needs to be explained is why and how it curves. If we have no idea about the mechanism, we are bound in a circle: curves produce gravity that results in curves.

It should be noted that the problem is not in the circle as such. It only reflects the helplessness of the "careless driver," who does not bother to find out what is under the bonnet of his car and thinks that it will go from point A to point B because the wheels will move the car and the car will move the wheels. A lack of understanding of the mechanism and its laws leads to the illusion of explanation in the form of a logical circle.

However, the concept of space and time as some entities out there and everywhere in the world is a result of an objectification error, which is also a logical fallacy. If we measure distances, sizes and positions of objects, then we take some reference samples ("rods") and reference points (zeros of a coordinate system). We use as many axes of the coordinate system as we want. We can take three axes and it will create the usual volume that our brain produces using the stereoscopic vision. It is necessary for accurate position, size and distance measurement. That is why the natural evolution of senses went from one-eye systems with flat two-dimensional vision to two eyes. But the evolution of artificial technologies allows us to make multidimensional measurements and describe them with complex tensors. Anyway, it does not mean that the Universe has three or as many dimensions as we take in our measurements. The Universe is non-dimensional or infinitely dimensional in the sense that dimensions are only the product of our measurement process. But if we mistake our spatial measurements (dimensions) for a special object called "space," we produce a phantom with its own shape and other properties. We make the category error of objectification: we create an object out of a process. The same goes for the concept of time. If we measure the dynamics of the world, we use reference periodic samplers to cut the evolving process into pieces and put them on a time scale. Our brain contains such "clocks," so we measure the dynamics of the world naturally as other animals do (more on this in further volumes). We also use external natural periodic oscillators (the Sun or the Moon) to make sense of the processes and their dynamics. To make more detailed measures, we create artificial oscillators (clocks) that are frequency-coupled with natural ones. But if we think of these measurements as the attributes of some special entity called "time," we are into an objectification error.

Here is a typical description: "Space is a three-dimensional continuum containing positions and directions. In classical physics, physical space is often conceived in three linear dimensions. Modern physicists usually consider it, with time, to be part of a boundless four-dimensional continuum known as spacetime. The concept of space is considered to be of fundamental importance to an understanding of the physical universe. However, disagreement continues between philosophers over whether it is itself an entity, a relationship between entities, or part of a conceptual framework … In the 19th and 20th centuries mathematicians began to examine geometries that are non-Euclidean, in which space is conceived as curved, rather than flat. According to Albert Einstein's theory of general relativity, space around gravitational fields deviates from Euclidean space. Experimental tests of general relativity have confirmed that non-Euclidean geometries provide a better model for the shape of space" (Wikipedia "Space").

The author leaves the debate to idle philosophers and speaks about substantialism (one of the views on space and time which considers them as transcendent entities that are outside of everything and contain everything) as if it is the only correct way to think about the issue. The objectification error goes unnoticed. Thus, it turns out that if nonlinear measurements better describe objects and their dynamics, this means that the curved shape of a certain "space-time fabric" in which all these objects exist is confirmed. This is the typical logical circle: if geometry (from Greek "land measurement") is curved, then the dimensions (from Latin "to measure") are curved. But as measurements (space and time) are taken for entities, the circle is not obvious and the statements sound like proofs of the GTR.

Let's look into the experimental tests of GTR's postulate about the existence of a curved space-time fabric. There are three "classical tests": the perihelion precession of Mercury's orbit, the deflection of light by the Sun, and the gravitational redshift of light. As we have seen in another article cited at the beginning of this chapter, they are called the predictions of Einstein, and his theory is considered their ultimate explanation.

What is so special about Mercury? As we have discussed in the previous volume, Newton's model of gravity has many deficiencies. One of them is the "two-body problem." The model describes celestial bodies only as two-body systems: the central body and the body that orbits the center along an ellipse. This simplification is quite useful because it allows us to calculate trajectories using linear equations with an accuracy that has been sufficient for a long time. The task of calculating the result for the number of interacting elements of more than two does not have an analytical solution in the framework of Newtonian mechanics. Newton's calculations according to Kepler's laws and their refinements, are the so-called solutions of the "unperturbed problem." The masses of other planets, their mutual influences, and the entire interacting system are completely ignored. It is a zero approximation: parameters are estimated with the assumption that there are no influences of external perturbations. The ideal situation is considered when the planet moves around the Sun in an elliptical orbit, and there are only two interacting objects. The real situation is different, and all elements of the system interact, which creates a disturbance of elliptical orbits.

Even the Solar system which consists of not too many bodies is a complex system of interacting elements that influence each other. Thus, the trajectories are not as simple as the classical model described. The movement of a planet is determined by many factors and is a perturbed trajectory. This means that the points of its phase space do not repeat with ideal periodicity. This concerns both the orbital and rotational trajectories. Their change is called precession. The most prominent change is the floating points of the closest distance to the Sun (perihelion) and farthest (aphelion). It happens with all the planets but Mercury being the smallest one and the closest to the Sun is influenced more than the others.

Newton's formula made it possible to describe the trajectories at a certain level of accuracy. Still, when displacements of the orbits (not only Mercury's) were discovered in the 19th century, the crisis of the classical theory of gravitation came. The measurements disagreement with the model were constantly updated but the explanation was lacking. Some suggested the existence of another planet by analogy with the success of the Neptune discovery based on perturbations of the orbit of Uranus. No evidence of a planet was found. But that did not mean that all the known elements of the Solar system cannot influence Mercury. However, it was considered the failure of the classical theory, and a new one was needed.

Here we have to remember that Newton's model did not explain the mechanism but just described the observed. Einstein took the existing measurements and suggested that the deflections can be explained by the curving space-time. Thus, it sounded like a prediction and confirmation of the suggested mechanism. Both statements are false. It was not a prediction but a post hoc description using new math that could help make better calculations. The mechanism was not disclosed. It does not matter, whether we say the geometry of the planet's movement or the geometry of space is curved. We need to answer why it is curved. The explanation should be qualitative, not quantitative. The mechanism must explain the observed and measured quantities. Otherwise, we are stuck in a circle of explaining measurement results by the measurement result. Geometry explains geometry. The physical meaning is absent.

But here is how the story is described by the GTR propaganda: "In general relativity, this remaining precession, or change of orientation of the orbital ellipse within its orbital plane, is explained by gravitation being mediated by the curvature of spacetime. Einstein showed that general relativity agrees closely with the observed amount of perihelion shift. This was a powerful factor motivating the adoption of general relativity" (Wikipedia "Tests of general relativity").

Einstein compiled a system of geometric tensors and adjusted the solution so that the numbers coincided with the existing data. There is no surprise that the numbers coincided. There are many ways to reach the same number. For example, the number 10 can be obtained by an infinite set of mathematical operations. But in physics, the point is in the physical meaning of the mathematical solution. If it reflects the physical mechanism, it can be considered a physical model with explanatory and predictive power. If it just describes reality, it is a phenomenological model and an interpretation of the observed.

Here is how physicist Julian Barbour describes this story: "It is a striking fact that all the mathematics Einstein needed already existed. In fact, I believe it is significant that he did not have to invent any of it. In 1915, he was immediately able to show that, to the best accuracy astronomers could achieve at that time, his theory gave identical predictions to Newtonian gravity except for a very small correction to the motion of Mercury … For many years, the sole discrepancy in the observed motions of the planets had been precisely such a perihelion advance for Mercury of exactly that magnitude. All attempts to explain it had hitherto failed. Einstein's theory explained it straight off" (Barbour, 1999).

Here we see two serious substitutions of concepts. First, the author calls the ability of the model to describe the already known measurements by a mathematical equation a prediction. It is not a prediction; it is a post hoc description. Second, the author calls this mathematical description an explanation. It is not an explanation; it is a description of the observed motion by an equation. These substitutions are so common when it comes to GTR that many physicists do not even notice them.

But later on in his book, the author has to admit that, in fact, it was not an explanation and a prediction: "Space-time is a beautiful sculpture. What makes it beautiful is the way in which its parts are put together. The fact that one can paint coordinate lines on the finished product and measure distance on the sculpture between points on it labelled by the arbitrary coordinates clearly leaves the sculpture exactly the same. All this changing of coordinates is purely formal. It tells you nothing about the true rules that make the sculpture. Belatedly, Einstein came to see that his whole drive to achieve general covariance as a deep physical principle had no foundation in fact. It was just a formal mathematical necessity … He hurried to a shop called 'Mathematics'… Straight off the shelf, at a bargain price, he bought a wonderful device called the Ricci tensor. Three years later, after agonizing struggles, he learned how to turn the handles properly, and out popped the advance of Mercury's perihelion and the exact light deflection at eclipses. But it never entered his head to ask how the device actually worked" (Ibid).

Indeed, Einstein did not invent any of the math that he used. But what he invented was the 'principles' that contradicted physical sense and created phantoms described by the math. In Special Relativity instead of Newton's absolute space and time he made the speed of light a reference frame for everything. Thus, a physical variable that depends on relative distance and time measurements became a self-referential absolute of the Universe (for details see "Part One. Music of Matter"). In General Relativity he used the idea of Hermann Minkowski about the union of space and time as some transcendent entity and the idea of Carl Gauss and Bernhard Riemann that it is possible to allow coordinate lines to be curved to express the positions on a curved surface. Thus, combining two ideas he invented the curving space-time fabric as the stage for the observable macrocosmic interactions. Making the fabric of space-time dynamical instead of rigid only allows the calculations of its metric to be adjusted more flexibly to the observable phenomena but does not give the model a real explanatory and predictive power.

It probably entered Einstein's head "to ask how the device actually worked," but the problem is that the device he was thinking about was not a physical reality that could be tested and explained. This "beautiful sculpture" was an abstract phantom and by a formal mathematical necessity a "finished product," the existence of which was just postulated as a "principle." From a physical science perspective, this is an impasse. From a metaphysical perspective, this is a classical way to form a religious concept.

Barbour wrote: "I am not claiming that the description of space-time given by Einstein or Minkowski is wrong. Far from it — they got it right, but they described the finished product, and the complete story must also include the construction of the product" (Ibid).

That's the problem: they can't be wrong. Their model is irrefutable as it speaks about an intangible transcendent entity. But this metaphysical stance does not relieve the question about the "construction of the product." Newton stumbled upon it and used the usual metaphysical way out: he declared God as the creator of the fixed space and time background. Einstein could not be explicit like that as the times were different, and he preferred to ignore the question of creating a product. The explanation of the physical mechanism of interaction was replaced by a math description of the "beautiful sculpture." We can make icons or sculptures for our gods, and describe them with words or math symbols, but that does not make these abstractions physical.

Old religions did not require formulas. The new ones in the spirit of the time should contain mathematical justifications for dogmas. Here it is:

$$R_{\mu\nu} - \frac{R}{2} g_{\mu\nu} + \lambda g_{\mu\nu} = \frac{8\pi G}{c^4} T\mu\nu$$

This is called the gravitational field equation. The name comes from the analogy with Maxwell's electromagnetic field equations that relate charges and currents to the supposedly existing entity "field" that is responsible for their behavior. The right side of the equation relates energy-momentum, gravitation, and the speed of light. It includes an auxiliary term called gravitational constant that was introduced to make Newton's gravitational equation correspond to the actual measurements of the physical phenomena. Thus, it is not about the physical mechanism behind gravitation but about adjusting the description of reality to this reality. But at least we can say that this side of the equation describes the actual physical world.

What happens with the left side is another issue. It is a set of tensors describing the curvature of space-time fabric geometry. We can say that it describes the "body of god" that curves as he wishes to make things of the material world behave as they do. It is no surprise that the tensors that describe this body are so complex that there is no solution to the equation. Actually, the left side is a short version. The full one contains multidimensional tensors and sixteen more equations. But what does this complexity that cannot be used in any practical way hide? It hides the banal trick: the postulated entity and its properties cannot be confirmed or refuted. It takes the model outside of the scientific domain and puts it straight into the belief domain.

But the left phantom side and right physical side have to converge. After all, it is an equation and there is an equality sign between them. How did Einstein make his god interact with the material world? The trick is also banal: he introduced an auxiliary variable that could take any value. It was called lambda or the cosmological constant. Anything goes to make ends meet.

People are still baffled by the complexity of math of this "simple and beautiful theory" and think that if it is complicated then it is correct. The truth is the opposite. Even the propaganda has to admit it. Here is how the author of an article in Wikipedia starts the usual praise: "Today, Einstein's theory of relativity is used for all gravitational calculations where absolute precision is desired, although Newton's inverse-square law continues to be a useful and fairly accurate approximation" (Wikipedia "Gravity"). Further, we read about the actual state of affairs: "A major area of research is the discovery of exact solutions to the Einstein field equations. Solving these equations amounts to calculating a precise value for the metric tensor (which defines the curvature and geometry of spacetime) under certain physical conditions ... Today, there remain many important situations in which the Einstein field equations have not been solved. Chief among these is the two-body problem ... The situation gets even more complicated when considering the interactions of three or more massive bodies (the "n-body problem"), and some scientists suspect that the Einstein field equations will never be solved in this context" (Ibid).

Now, where is it used where the precision is desired if there are no precise solutions for the actual reality? The author tries to put a good face on a bad game, but we see that it is all a bluff. The outcome is that the Newtonian model is still used in practical applications. Why use complex tensors that do not reflect reality? It is better to make it simple and correct based on experience. Thus, practical physics remains an ad hoc game of chance and still waits for a model that will be able to make true predictions not post hoc adjustments using math tricks.

Even with the help of the almighty curving space-time fabric, GTR cannot describe the actual interaction and says nothing about the physical mechanism behind it. Thus, it cannot explain the observed interaction results that speak of attraction, repulsion and balanced state. "Why matter clumps together into galaxies of a certain characteristic range of sizes, rather than either dispersing completely or massing into a single superclump ... Einstein's GR equations are unstable, requiring a "cosmological constant" (i.e. fudge)" (Hotson, 2002). Even the name is a fudge: it is not a constant, but a variable that can be attributed with any value.

The propaganda calls GTR "the most beautiful of all existing physical theories" and "the simplest theory consistent with experimental data" (Wikipedia "General Relativity"). The truth is exactly the opposite: its ugly complexity does not allow for practical solutions and its math descriptions are internally inconsistent and use arbitrary variables to hide contradictions. As for the consistency with experimental data, we have seen in the previous volume that the actual situation is again the opposite and we will show more examples in this part of the study. The aim is not to refute GTR. After all, it is not a physical theory but a belief system that employs immaterial entities so it cannot be confirmed and refuted. The aim is to show what the actual experiments tell us and to explain them within the proposed model.

It is interesting to note that even the dirty trick of putting an arbitrary variable into an equation is praised by the propaganda as the start of modern cosmology. Here is the true story. The mathematician Alexander Friedmann realized that there are different possible outcomes for the Universe as prescribed by Einstein's equation: the initial zero radius with its expansion and the oscillation between growth from zero to expansion and subsequent narrowing, and so on in a cycle. This is where the Big Bang theory or the Standard Model of Cosmology (SMC) comes from. It is just a result of mathematical speculations about the possible geometry of the phantom fabric using any possible value for an arbitrary variable. Einstein even resisted this interpretation of his equation as he thought of the Universe as a stationary "bubble" with a constant radius. He sent an angry letter to the journal that published Friedmann's article saying that the calculations did not make sense and were based on math errors. Friedmann showed that there was no mistake. Einstein resisted for a long time but finally admitted that his opponent was right since mathematics was obvious.

The author of the review article dedicated to the centennial of the GTR and published in one of the leading scientific journals in the world with the corresponding title "Science" wrote: "Einstein allegedly came to regard this amendment as his "greatest blunder," but from a modern perspective, it is looking more like his most extraordinary insight … The theory of general relativity (including λ) provides a singular prescription for describing classical gravity and nongravitational physics in this environment. There seems to be no need for baroque variations and empirical corrections over its domain of applicability. It has grown from an essentially intuitive quest to providing the foundation for high-energy astrophysics and cosmology, in which it is routinely considered to be correct" (Blandford, 2015).

What is a theory that does not require empirical corrections and is the only prescription for describing the world? It is a religious dogma, which has nothing to do with science. However, the article is published in a scientific journal. It is the result of the propaganda of the dogmatic "holy scripture" of the prophet. Einstein even formulated his hypotheses as postulates, that is, as eternal commandments carved on stone tablets. So, there was no blunder. Prophets do not make mistakes; they only have revelations. Their postulates cannot be refuted. This is why an auxiliary variable was introduced into the equation in the first place. If necessary, one can always say that the variable is zero, and with a different version of events, give it any value. Thanks to this, subsequent apologetics can do anything with this variable, as long as the dogmas remain intact. Lambda has turned into a "Cheshire cat," which is either nullified or given some value, then simply ignored or left in the form of a smile just in case the cat comes in handy.

Here is a confession of a physicist who used to be one of the believers: "I will speak of conjectures that were widely believed to be true, in spite of never having been proved. But I was among the believers, and I made choices about my research based on those beliefs. I will speak of the pressures that young scientists feel to pursue topics sanctioned by the mainstream in order to have a decent career. I have felt those pressures myself, and there were times when I let my career be guided by them. The conflict between the need to make scientific judgments independently and make them in a way that doesn't alienate you from the mainstream is one that I, too, have experienced … But in both quantum theory and general relativity, we encounter predictions of physically sensible quantities becoming infinite. This is likely the way that nature punishes impudent theorists who dare to break her unity. General relativity has a problem with infinities because inside a black hole the density of matter and the strength of the gravitational field quickly become infinite … At the point at which the density becomes infinite, the equations of general relativity break down. Some people interpret this as time stopping, but a more sober view is that the theory is just inadequate" (Smolin, 2006).

The propaganda sounds like this: the theory provides a singular prescription for describing classical gravity and nongravitational physics, i.e., the whole world. The truth sounds like this: the theory is not adequate to reality. Why do millions of physicists and billions of laymen consider it to be the most beautiful and correct theory? The first turned off their critical thinking and became believers; the second just followed the propaganda of the mainstream that is voiced from all possible media sources.

Here is a typical example of such propaganda. The Standard Cosmological Model is derived from the GTR using one of the versions of the cosmological constant — a small positive value. Thus, the equation solution leads to an expanding Universe. The curving but static in its boundaries fabric of Einstein's model turned into a curving and stretching fabric. But we should remember that Einstein's tensors are about the space-time geometry not about a physical Universe. The mainstream is so confused with the concept of this entity that it mixes it with the physical reality and gets even more confused.

Let's read the attempt at an explanation: "The expansion of the universe is the increase in distance between any two given gravitationally unbound parts of the observable universe with time. It is an intrinsic expansion whereby the scale of space itself changes. The universe does not expand "into" anything and does not require space to exist "outside" it. Technically, neither space nor objects in space move. Instead, it is the metric governing the size and geometry of spacetime itself that changes in scale … The metric expansion of space is of a kind completely different from the expansions and explosions seen in daily life" (Wikipedia, "Expansion of the Universe").

It is impossible to make any physical sense from this description. Is it an increase in distance between parts, or is it not? Nothing is moving, neither space (some kind of entity-container for objects) nor objects. Nothing is expanding into anything or outside of anything. Then what is expanding? The metric of space-time. The space does not expand, but its scale does. How come? Well, it's a special kind of expansion that you do not see in daily life, and your common sense (physical experience) cannot comprehend it. What a beautiful and consistent description. Such mess is at the heart of the modern cosmological theory that pretends to be the standard one, i.e., the only possible one.

Here is the praise: "The model assumes that general relativity is the correct theory of gravity on cosmological scales. It emerged in the late 1990s as a concordance cosmology, after a period of time when disparate observed properties of the universe appeared mutually inconsistent, and there was no consensus on the makeup of the energy density of the universe" (Wikipedia "Lambda-CDM model").

So, the new theory is based on the correct old theory which is actually inadequate to reality and mathematically inconsistent but helps put the observed properties into a consistent whole. What are these properties? The main one, on which all the structure holds, is the alleged expansion of the Universe. This hypothesis which was never proved has turned into an axiom. If we ask an average educated person about it, the answer will sound like this: science has long ago discovered that the stars are running away from us, which means that the Universe is expanding. We can also read it in the mainstream mouthpiece: "The observation in physical cosmology that galaxies are moving away from the Earth ... The velocity of the galaxies has been determined by their redshift, a shift of the light they emit to the red end of the spectrum" (Wikipedia, "Hubble's Law").

We get even more confused: does expansion mean the movement of objects or is it a change in the scale of space-time? One article states that nothing is moving anywhere, and another that it is moving and even gives the direction of movement away from the Earth. Meanwhile, both describe the same cosmological model. Don't try to find any consistency because there is none. Theorists who consider GTR to be the correct theory are themselves confused. Of course, they are trying to find some physical meaning and think about moving objects, and not about some ghostly extension of the phantom metric. But has science discovered that stars are running away from us? No. It discovered the redshift of the light coming from the stars and reaching us. However, this observation in any textbook or encyclopedia is interpreted as the expansion of the Universe. This is a typical example of a substitution of concepts. One possible explanation became the observation that "galaxies are moving away from Earth."

But if we take this assumption as a basis, it turns out that it not only does not provide answers to old questions about gravity, but also gives rise to new ones. Why are distant galaxies moving away from Earth? Are we really the "navel" of the Universe? Why do galaxies move away but do not disintegrate as structures? After all, if the fabric of space-time is omnipresent and expands, then everything should fly apart in all directions. If gravity is an illusion, and everything moves in the curves of this stretching fabric, then why are there stable planetary structures in the observable space? Moreover, the galaxies themselves retain their structure. Not to mention the fact that objects on Earth do not fly apart into small pieces unless they undergo expansions and explosions "seen in daily life." The answer from believers in the phantom but omnipresent and omnipotent fabric of space-time is one: the ways of this thing are not inscrutable. Even the prophet did not say anything about "how the device actually worked."

We need to return the physical meaning. Edwin Hubble, after whom the "law of expansion" is named, discovered the redshift, not the expansion of the Universe. This is not the same thing, since different explanations for the observed spectrum are possible, and the movement of distant stars even further away from us into nowhere is only one of them, and not the most plausible.

Let's start in order. What is redshift physically? If we proceed from the normal physics of waves in a medium, then everything is simple: this is an increase in the length (decrease in frequency) of a light wave and a transition to the range that our brain perceives as red. Thus, we are talking exclusively about the observer measuring the parameters of the electromagnetic waves arriving at him. This is the process of relating the spatial and temporal dimensions of light signals. The same thing happens with any environmental signals in other energy ranges. Our brain does this with its natural tools (more details in subsequent parts of the study). We have also created artificial instruments that help us make more accurate measurements. For example, what the brain perceives as the red color of different shades, we can decompose into spectrum components and express them in hertz (the number of vibrations per second).

So, we are talking about the process of measuring dependent variables of wave processes, and not about the properties of some phantom fabric that has its own metric. Let's remember the basics: the parameters of a wave depend on the properties of the source of vibration, the propagation medium, the receiver, and the distance between the source and the receiver. If these parameters do not change, then the wave will have a certain length. If they change, then the length also changes. If the source and receiver do not change, and the distance between them changes, then the length will change for a simple reason: the waves arriving at the observer (meter) will arrive faster and more often if the source is approaching, but slower and less frequently if the source is moving away (Doppler effect). In other words, the source may oscillate at the same frequency, but the perceived frequency varies. This once again proves the Animacentric Principle of Relativity (APR), proposed within the framework of TEH and which we discussed in the previous volume. Thus, the siren of an approaching ambulance will first increase in frequency for the observer (listener) and decrease as it moves away. The same goes for light. If the frequency increases, the light will shift towards the shorter wavelength spectrum (blueshift). In the case of a decrease, it goes towards the long-wave spectrum (redshift). However, since the speed of light waves is enormous, any change can be noticed only if the speed of movement of the source and the observer (viewer) relative to each other and the change in distance are commensurate with the speed of light. However, we observe different spectra of light waves precisely because there are other factors besides distance. The key one is the wave propagation medium. Red and blue shifts are possible not only and not so much due to the approaching and moving away of the source.

That is, the version about the objects in cosmic distances moving away from us and, as a consequence, the presence of a redshift is a working hypothesis, but not the only one. However, it completely fits into the wave model of light. It speaks about the movement of objects sending and receiving waves in the medium. But GTR speaks about the curvature, stretching or compression (depending on the value of an arbitrary lambda variable) of the space-time fabric. Light, in the framework of Einstein's model, is not waves, but the flight of massless photons through emptiness with a constant and absolute speed. It's not about waves. The Doppler effect is not compatible with it in principle. What if the observed reality does not correspond to the model? Should we change the model? No way, believers in its postulates tell us. We just need to come up with a special redshift that has nothing to do with "daily life" wave phenomena: cosmological redshift.

Let's hear the mouthpiece of the mainstream: "There is a distinction between a redshift in cosmological context as compared to that witnessed when nearby objects exhibit a local Doppler-effect redshift ... One interpretation of this effect is the idea that space itself is expanding ... However, expanding space is only a choice of coordinates and thus cannot have physical consequences. The cosmological redshift is more naturally interpreted as a Doppler shift arising due to the recession of distant objects ... The resulting situation can be illustrated by the Expanding Rubber Sheet Universe, a common cosmological analogy used to describe the expansion of space. If two objects are represented by ball bearings and spacetime by a stretching rubber sheet, the Doppler effect is caused by rolling the balls across the sheet to create peculiar motion. The cosmological redshift occurs when the ball bearings are stuck to the sheet and the sheet is stretched ... Describing the cosmological expansion origin of redshift, cosmologist Edward Robert Harrison said, "Light leaves a galaxy, which is stationary in its local region of space, and is eventually received by observers who are stationary in their own local region of space. Between the galaxy and the observer, light travels through vast regions of expanding space. As a result, all wavelengths of the light are stretched by the expansion of space. It is as simple as that" (Wikipedia, "Redshift").

It is as simple as that. Nothing is moving anywhere but space-time fabric stretches. However, this is only a geometrical metric that does not have any physical consequences. So, the redshift should be naturally interpreted as the movement of objects. But they do not move and only the "rubber sheet" of the Universe stretches leading to a peculiar motion which is not motion. Everything is stuck to it but it is not a physical entity as it is only a choice of coordinates (measurement process).

But how can something intangible stretch and lead to the movement of material objects? Not in physics. In religion, such an immaterial entity can do whatever it wants. This is the difference between a real redshift due to the movement of objects and a cosmological one, caused by the expansion of an immaterial entity and not related to the movement of objects. Cosmological redshift is a phantom caused by the phantom of the fabric of spacetime. This is why the entire quote is filled with contradictions. The author of the article is struggling with cognitive dissonance when it comes to describing something physical using concepts that "cannot have physical consequences." The quoted cosmologist tries to pretend that he has overcome this dissonance and talks about the expansion of space, as if it were the actual stretching of a certain elastic band. The error of objectification, which takes measurements of changing coordinates for a really existing "rubber sheet" of the fabric of space-time, and the lack of critical perception of the dogmas of GTR leads to the conclusion that "it is as simple as that," when, on the contrary, everything is very confusing if we try to find at least some physical meaning. This is why the mainstream prefers to turn off critical thinking and believe.

The simple analogy with the balls glued to a rubber sheet does not work for one more reason. If space-time is a kind of omnipresent fabric, then it must contain all objects, including the planets closest to us, the Earth, ourselves, and all the elements of the atomic and subatomic world. If such an object begins to stretch, everything should fly apart in all directions, not just some distant stars. If the space to which we are "glued" is expanding, then we must tear up into small pieces and demonstrate the cosmological redshift. Is something like this happening? No. Logic dictates that the premise about the stretching space-time fabric is wrong. However, believers do not despond and say that the cosmological redshift is cosmological exactly for the reason that it has no concern about every little thing like planets, human beings and, especially, atoms. But in Einstein's model, the fabric of space-time is an all-encompassing background and makes no exception for the level of matter. This was the main dilemma of the prophet, which bothered him until the end of his life: he could not "glue" his invented fabric to the microcosmic level and explain the phenomena occurring in it with the help of this new god, who turned out to be omnipresent, but not omnipotent. However, it is useless to expect consistent logic where there is none in the first place.

If we get down to "daily life" physical meaning we understand that there are several options for explaining the observed redshift and not a single one needs magical tricks from the hat that produces phantom entities that expand into nothing in the way that we cannot imagine and produce peculiar motion without movement.

It may be the result of the usual Doppler effect due to an increase in the distance between the wave source and the receiver (observer) for a reason that is banal from the point of view of wave theory. As the source of a wave moves away from the observer, each successive wave crest comes from a greater distance, so the wave takes longer to reach the receiver. An increase in the period of registration of ridges by the observer-receiver means an increase in the period and a decrease in frequency (inverse variables). Everything is logical, understandable and within the framework of our "daily life" experience. But again, this is just a hypothesis that the redshift occurs due to the stars moving away from us. In other words, it is not a fact that the Universe is somehow expanding in all directions. The fact is that the radiation spectrum shifts to longer waves.

If we assume that light is waves in a medium, then there may be another phenomenon that is quite commonplace for wave behavior. The redshift may simply result from the loss of energy by the wave due to dissipation in the medium. Accordingly, the radiation can go into the low-frequency range, and shift in the visible spectrum to red. It is such a natural explanation that at the beginning, both astronomers and physicists perceived the detected redshift in this way and did not make their eyes wide with horror: "The Universe is scattering to nowhere!", "The end of the world is inevitable!"

The interpretation of redshift as a result of dissipation in the medium is consistent with the second law of thermodynamics, does not contradict the energy laws and physical causality, is subject to empirical testing, and has explanatory and predictive power. This hypothesis is natural, and it existed even before the discovery of the redshift. Some physicists predicted that light from distant stars would inevitably demonstrate energy loss. When the redshift was discovered, it was perceived as confirmation of such a prediction. Data on the parameters of the change in the emission spectrum of galaxies showed that energy losses are proportional to distance. It once again proved that light is a real physical wave in a physical medium and not a flight of angelic photons in a void.

But now you will not find the version about the redshift of the spectrum of galaxies as a result of the dissipation of the energy of the light wave in any description within the mainstream. How did a physically plausible explanation fall out of sight? The "blinkers" of the dogma were put onto the eyes of the believers. Oxymorons "expanding empty space" and "curving empty space" became the basis of the cosmological theory, which is considered to be leading in modern science. When rare daredevils, who have not traded common sense for academic titles and laurels of adherents of the "singular prescription," ask, why not assume that there is a natural dissipation of energy during the propagation of light from stars, disciples ask a counter-question: where will energy dissipate if there is a void? The logical circle closes on the void.

Edwin Hubble considered various interpretations of the observed redshift phenomenon. He called the expansion hypothesis a "forced interpretation of the observational results" (Hubble, 1936). But by that time, Einstein was running the show of theoretical physics. It often happens in science: the choice in favor of interpretation depends on the model that is popular at the time of choice, and then everyone for a long time forgets that another option is possible at all. As a result, a paradox arises: the empirical fact of redshift, which can be considered as evidence of the presence of a medium and energy dissipation, is interpreted as proof of the correctness of both STR and GTR, which actually do not predict it and even contradict it. One more paradox of the history of science: the "law of expansion" and the "constant of expansion" are named after Hubble, although the scientist himself was against this hypothesis.

But there is another physically plausible version of the explanation. Redshift can be an optical effect: a shift towards the background. For example, it happens on Earth due to the influence of the atmosphere, so the effect is called the aerial perspective. Distant objects (for example, a chain of mountains against the sky) shift to the blue. The redshift is also possible if you look at the perspective in the rays of sunrise and sunset: objects will be painted in the background spectrum.

There is one objection to the interpretation of redshift as a background effect of the medium: there is no such effect in the dark. Hubble calculated the redshift based on an analysis of the frequencies of the stars' spectral lines compared to the frequencies of the corresponding lines recorded in the reference measurements in laboratory darkness. But who said that there is reference darkness where the telescopes are looking? Maybe there is still a kind of "aerial" perspective of background? The inverted commas show that what is meant is not the presence of air, as a set of gases, but the presence of any material medium in principle.

But the most important thing: it does not matter which version of the redshift explanation we take as the basis. They are all related to the propagation of physical waves in the physical medium and the dependence of their characteristics on the parameters of the source, receiver, their relative position and the parameters of the propagation medium. And only the "singular prescription," which is "routinely considered to be correct" and "does not require empirical corrections," speaks of the flight of virtual angels in a phantom void with constant and absolute speed.

We dealt with two of the three "classical tests" of the predictions of the GTR. We saw that the perihelion precession of Mercury's orbit was not a prediction of GTR but a previously known fact that could be described with the geometrical tensors used by Einstein. The description of geometrical trajectory by geometrical equations is only natural but it does not explain why the trajectory is like this. Moreover, Einstein's equations do not solve even the mathematical riddle of multiparameter systems with many interacting bodies and, accordingly, cannot be called solutions of the actual physical observation. The prophet understood this well so he just chose to hide behind the circular reasoning: the geometry of the planet's movement is due to the geometry of space-time fabric. Billions of people are used to the circular logic of religious concepts that in essence goes like this: things exist because God exists and makes them, so if things exist, then God exists. Thus, those who believed in the new transcendent entity (space-time fabric) proclaimed by the prophet of the new religion took this "explanation" without any doubt and still called the precession the greatest victory of the greatest theory.

The same happened with the redshift. Instead of thinking about physical reasons for the observed, the "explanation" by the expanding metric of space-time was taken for another marvelous prediction and glorious triumph of the new religion. The believers did not even notice that redshift contradicts both STR and GTR postulates. The prophet himself admitted this was his biggest blunder, but for the parishioners of his church this was no longer important: prophets do not make mistakes, but only outstanding revelations that do not need empirical corrections.

But what about the deflection of light by the Sun or "gravitational lensing" as they call it within the GTR church? It is easy to show that it is not what they say it is. First, the description of an observation is not a prediction because if the mechanism is not disclosed the model does not have an explanatory and predictive power. What the believers call a prediction of GTR was again just a better description made by Einstein using the complex tensor geometry which he finally could handle. As Julian Barbour noted: "After agonizing struggles, he learned how to turn the handles properly … But it never entered his head to ask how the device actually worked" (Barbour, 1999). It also did not occur to the believers to ask the prophet what the mechanism of deflection was. They habitually took the circular logical fallacy for an answer: light bends because space curves.

The story completely coincided with the redshift story where the normal wave phenomena were treated as some magical happenings with space-time in the void and called cosmological redshift. The wave theory has no problems with explanation. Refraction happens due to a change in speed and length of the waves that depend upon the properties of the medium. If they change, the wave path changes. We can observe it in water waves passing from deep to shallow water, slowing down, decreasing in length, and bending their direction. We can observe it with light passing through air with different atmospheric conditions or from air to water. Waves naturally bend as they pass around the edge of an object. Again, the trajectory depends upon the parameters of the wave and the medium. For example, when light bends around droplets of water in the clouds we can observe fringes of light, dark or colored bands, or coronas surrounding the Sun or the Moon. The intuitive explanation comes again from the hydrodynamic analogy: if we imagine light waves as water waves, we understand that the diffraction is the result of constructive or destructive interference as frequency-phase interaction of waves. If the crests of the waves combine, they are amplified to a certain value and light will appear brighter. If the crest of one wave meets with the trough of another wave, they cancel each other out by a certain value and we see less bright light or even darkness. Various frequency-phase combinations can cause different interference patterns.

These are basic facts of wave life. They were, of course, known to Einstein and his followers. What was the problem? The void was the only problem. Do you remember the first step he had to take? "The only way to arrive at a satisfactory theory is to give up the notion of a medium filling all the space. This is the first step to be taken" (Einstein, 1910). What was left? "The region of space without matter and without an electric field appears completely empty, i.e., it cannot be characterized by any physical quantities" (Einstein, 1918). Next was a step backward to the Newtonian corpuscular model of light. Waves are the propagation of oscillations in a medium. If there is no medium, there can be no waves. If there is emptiness, what can light be? The flight of particles through the void. What can alter their magical flight path? Here we do not even concern ourselves with the question of how this flight is supported throughout this long journey if there is nothing in between the sender and receiver. Newton just thought that the particles were lucky not to be hindered by anything in the void. But massive celestial bodies are an obvious obstacle. It would seem that gravitation is a natural explanation. If they are attracted by a massive body, they may change their path. This was predicted by Newton. The problem was to calculate the value of the deflection as the mass of such particles was unknown and it was supposed to be so tiny that calculations could not produce any notable result. The theme was dropped.

When wave theory of light came with better explanations and explained what corpuscular theory could not explain at all, the "gravitational lensing" was not even thought of. But by the end of the 19th century when mechanical models of ether faced a dead-end the theme was resurrected. At the beginning of the 20th century, Einstein canceled the medium for light waves and called light the flight of massless particles (photons) in the void. But the old prediction that light should bend around massive objects remained. The dilemma arose: how can a massless entity take part in a gravitational interaction the strength of which is defined by the masses of the interacting bodies? How can an intangible angel of the new religion be influenced by some material thing? The dilemma seems unsolvable.

As Barbour noted, Einstein "got into a muddle — but a most creative muddle" (Barbour, 1999). Nothing could stop him from creating a powerful virtual entity to explain the behavior of tiny virtual entities. Space-time fabric "explains" it all. It just curves and makes light bend. So, when Arthur Eddington, Frank Dyson and their collaborators in the experiment during the solar eclipse in 1919 made a photo of light slightly bending when passing near the Sun, it made the front pages of most major newspapers and the believers cheered the prophet saying that his theory was confirmed.

Einstein was so proud of the success that he put himself above God. When asked by his assistant what his reaction would have been if general relativity had not been confirmed by Eddington and Dyson, Einstein said "Then I would feel sorry for the dear Lord. The theory is correct anyway" (Rosenthal-Schneider, 1980). Of course, he could put God of the old religions aside, as he invented a new one, an intangible Space-Time Fabric, that was also irrefutable to any test. We can put it the other way: happenings in the world were confirmations of its existence. The theory is correct anyway.

This is the story of how normal physical wave deflection in a medium and around an obstacle became magical space-time curvature influencing the flight of massless angels in the void. Physics was traded for metaphysics and the audience applauded as it was used to religions talking of miracles to explain reality. Even the term "gravitational lensing" that was coined was a typical fudge: GTR did not speak of any lensing (optical wave effect) or any gravitational (attracting) force. It spoke only of space-time metrics.

From this glorious start of a new religion with the three "classical tests" that were not tests at all, other miracles of this "most creative muddle" followed. For example, propaganda calls gravitational singularities or black holes one of the greatest predictions and confirmations of theory. For a person who holds to his common sense and remembers what mathematical consistency means, any solution of the equations that leads to infinities means that something is wrong with the model. We can repeat the quote from Lee Smolin: "General relativity has a problem with infinities because inside a black hole the density of matter and the strength of the gravitational field quickly become infinite … At the point at which the density becomes infinite, the equations of general relativity break down. Some people interpret this as time stopping, but a more sober view is that the theory is just inadequate" (Smolin, 2006).

This is indeed a sober view. But a "drunken" view about the space-time fabric is the following: it can curve as it wishes and when it curls up so tightly that it becomes infinitely dense a black hole appears that swallows everything that is unlucky to be nearby into the gravitational collapse from which nothing can escape. If it sounds like a mockery, it is. But it is not a mockery about the "greatest theory" but a description of how this theory mocks physical and common sense. Here is the definition of a black hole: "A black hole is a region of spacetime where gravity is so strong that nothing, including light or other electromagnetic waves, has enough energy to escape it ... The boundary of no escape is called the event horizon. Although it has a great effect on the fate and circumstances of an object crossing it, it has no locally detectable features according to general relativity" (Wikipedia "Black Hole").

If anyone has doubts that GTR is not about physics but metaphysics, this definition should clear everything. It proclaims the existence of something that cannot be found, thus it produces a postulate that cannot be verified or refuted by an empirical test. This is a religious domain, not a scientific one. A question from an audience of this magical show: how can we verify that the rabbit appeared from the hat if no signals come from it? Once again: nothing can escape from this object (or whatever it is), so we cannot get any signal to confirm that it exists. On the other hand, we cannot refute that it doesn't. As with any religion, the trick is obvious: the proclaimed entity is an object of belief. We are offered two alternatives: either we believe that it exists or we believe that it does not. Both are not about science. The dogma says that this entity has a great effect on the material world but it has no detectable features. Isn't it the standard definition of a god?

Actually, Einstein had nothing to do with black holes. He only produced equations of space-time metric that led to infinite density solutions (singularities). It was the believers and followers of the prophet who proceeded to create phantoms from this muddle. Here is how one of them, Steven Hawking, described the story: "The universe could have had a singularity, a big bang, if the general theory of relativity was correct. However, it did not resolve the crucial question: Does general relativity predict that our universe should have had a big bang, a beginning of time? The answer to this came out of a completely different approach introduced by a British mathematician and physicist, Roger Penrose, in 1965. Using the way light cones behave in general relativity, together with the fact that gravity is always attractive, he showed that a star collapsing under its own gravity is trapped in a region whose surface eventually shrinks to zero size. And, since the surface of the region shrinks to zero, so too must its volume. All the matter in the star will be compressed into a region of zero volume, so the density of matter and the curvature of space-time become infinite. In other words, one has a singularity contained within a region of space-time known as a black hole" (Hawking, 1988).

The GTR did not predict anything but anything could be derived from its equations. Of course, gravity is about attraction. After all, this is what the term means. But the fundamental interaction that Newton called gravitation just because "apples fall on the heads" is not only about attraction. Otherwise, everything, including all apples, would collapse into a black hole with zero size (whatever that means). This problem of gravitational collapse that the model predicted but was not observed in reality worried Newton and other physicists of the classical era. As the model did not disclose the mechanism of interaction, it failed to account for all its manifestations. In other words, it was inadequate. However, the adherents of the new religion were not bothered with reality. For them everything was simple: if equations speak of a one-way direction, then gravitational collapse should be real. Mathematicians created entities or, to be precise, phantoms. Now what should they do if the observed reality refuses to abide by the postulates of the prophet? The only way out is to invent some other reality behind "the event horizon" which we cannot observe but it exists just because equations say so.

For a sober view zero size or volume is just nothing and cannot be part of a physical theory that should be dealing with material things just by definition of science. But for a believer who is mesmerized by religious dogmas, it exists but hides from our view. It would seem that "general relativity, by predicting points of infinite density, predicts its own downfall" (Ibid). But no, it robs its adherents of common sense and they go on in search of nothing. Why? Hawking has the answer: "We believe that there are much larger objects in the universe, like the central regions of galaxies, that can also undergo gravitational collapse to produce black holes" (Ibid). The key word is "believe."

But how can anyone search for something that is nothing? Here is the consolation: "The singularities produced by gravitational collapse occur only in places, like black holes, where they are decently hidden from outside view by an event horizon. Strictly, this is what is known as the weak cosmic censorship hypothesis: it protects observers who remain outside the black hole from the consequences of the breakdown of predictability that occurs at the singularity" (Ibid). This "cosmic censorship" hypothesis offered by Roger Penrose postulates that we cannot find the black hole from the outside because, as Hawking puts it, "God abhors a naked singularity." Theoretical physicists who follow Einstein's steps cannot but come to God as the ultimate explanation of everything. They pretend that they use this concept as a metaphor but the truth is that there is no other way for them: if their theory speaks of all sorts of intangible entities in the void (nothing in nothing), God is the only consolation.

So, the phantom of a black hole is hiding from us beyond the event horizon and we cannot witness its kingdom come, but when it comes there will be no difference for us as we will be beyond the event horizon too. Does it remind you of something? Nothing is new under the Moon (or under the black hole?). All religions have this narrative. Hawking cannot escape the association himself: "One could well say of the event horizon what the poet Dante said of the entrance to Hell: "All hope abandon, ye who enter here." Anything or anyone who falls through the event horizon will soon reach the region of infinite density and the end of time … Thus, in a sense, we are still all doomed, even if we keep away from black holes" (Ibid).

Nothing can save us from the black holes that live in the minds of believers in GTR. This is what they called physical theories in the 20th century and continue to call to this day. Moreover, those who look for this something that is nothing beyond anything at the end of time call themselves scientists and even get prizes in physics not in fiction. Penrose got the Nobel Prize in Physics "for the discovery that black hole formation is a robust prediction of the general theory of relativity" (The Nobel Prize in Physics 2020. NobelPrize.org). To make it sound as if it is actually about physics and has an empirical base, the committee gave the second half of the prize to Reinhard Genzel and Andrea Ghez "for the discovery of a supermassive compact object at the centre of our galaxy" (Ibid). Of course, it did not have anything to do with Penrose's theory as per definition within his model you cannot discover a black hole. Moreover, the statement that astronomers discovered some object is false. What they observed was the rotation of the stars around a center without any object in it. But the interpretation of this finding within the mainstream dogma was straightforward: "According to the current theory of gravity, there is only one candidate — a supermassive black hole" (Ibid).

It is interesting to note that Hawking struggled with the cognitive dissonance that was produced by such a search for a phantom (gravitational singularity) born from a phantom (curving space-time). For many years he was trying to find the way out of a conundrum that can be expressed in a simple question: "How could we hope to detect a black hole, as by its very definition it does not emit any light? It might seem a bit like looking for a black cat in a coal cellar" (Ibid). More precisely, it is like looking for a black cat that does not exist. Mission: impossible.

At first, his consolation was that astronomers observe systems of stars that orbit around some point in space that does not emit anything. Obviously, this was not good enough as there could be many reasons for this. Hawking was thinking about an unseen faint star as he was in grips of a one-way view of the gravitational interaction. But if we think of any fundamental interaction as a two-way street that has attraction and repulsion depending upon the parameters of the interacting elements, the most obvious answer is that the stars rotate around some center that may not be any object at all but just a point of balance. But for Hawking it was out of the scope of his religion: "There are other models to explain (such systems) that do not include a black hole, but they are all rather far-fetched. A black hole seems to be the only really natural explanation of the observations" (Ibid). For a believer, a hypothesis about nothing with zero size beyond the event horizon is a natural explanation but a hypothesis that does not involve such a phantom is far-fetched.

Hawking was sure that black holes existed as he believed in them because "a theory was developed in great detail as a mathematical model before there was any evidence from observations that it was correct" (Ibid). It is not a problem if some model predicts something that can be observed but has not been observed yet. The problem starts when the model predicts something that is nothing and cannot be observed. But the believers tend to find "confirmations." Here the story is the same as with other tests of GTR. All experimental facts that spoke of real wave processes were interpreted as proofs about the correctness of the model that contradicted them.

Hawking wrote about the redshift from a massive object discovered in 1963 and began to contradict his own notion of a black hole: "The only mechanism that people could think of that would produce such large quantities of energy seemed to be the gravitational collapse not just of a star but of a whole central region of a galaxy" (Ibid).

How can a gravitational collapse emit energy and show the red-shift if by definition nothing escapes from it? But the author insists: "Further encouragement for the existence of black holes came in 1967 with the discovery ... of objects in the sky that were emitting regular pulses of radio waves" (Ibid). How can radio waves be the confirmation of black holes? Logic fails the believers when they seek for a black cat that does not exist. But the belief does not evaporate: "If a star could collapse to such a small size, it is not unreasonable to expect that other stars could collapse to even smaller size and become black holes" (Ibid).

But nothing helped to overcome the inherent contradiction: the need to find something that cannot be found by the definition given by the seekers themselves. Hawking made the only possible choice: he changed the definition. Now black holes could emit radiation but only if quantum effects were taken into account. This was called "Hawking radiation." To be precise he canceled the singularities that Penrose calculated from Einstein's equations.

The revelation came as light emitted from a star: "This can produce some remarkable consequences, such as black holes not being black, and the universe not having any singularities" (Ibid). But it contradicted the great mathematical model that he was sure was a correct one. To escape from such a whirl of dissonances the GTR model had to be married with the quantum world that was full of its own dissonances.

The new version was that black holes actually emitted particles. Here all the usual tricks of the quantum magical show began: "How is it possible that a black hole appears to emit particles when we know that nothing can escape from within its event horizon? The answer, quantum theory tells us, is that the particles do not come from within the black hole, but from the "empty" space just outside the black hole's event horizon! ... These particles are virtual particles like the particles that carry the gravitational force of the sun: unlike real particles, they cannot be observed directly with a particle detector ... The idea of radiation from black holes was the first example of a prediction that depended in an essential way on both the great theories of this century, general relativity and quantum mechanics. It aroused a lot of opposition initially because it upset the existing viewpoint: "How can a black hole emit anything?" When I first announced the results of my calculations ..., I was greeted with general incredulity ... The existence of radiation from black holes seems to imply that gravitational collapse is not as final and irreversible as we once thought" (Ibid).

Of course, people were skeptical at first. They were getting ready for the doomsday but now it was called off. But then a great relief came to the parishioners. We are not doomed after all and black holes will not carry us beyond the event horizon from which no one can escape! We should not abandon all hope. The savior has come and canceled the collapse or made it reversible. The virtual angels of mercy flying on their wings through the void will bring the glad tidings. How can they perform this feat?

"The explanation of how black holes can emit particles and radiation was that one member of a virtual particle/antiparticle pair (say, the antiparticle) might fall into the black hole, leaving the other member without a partner with which to annihilate. The forsaken particle might fall into the hole as well, but it might also escape from the vicinity of the black hole. If so, to an observer at a distance it would appear to be a particle emitted by the black hole. One can, however, have a different but equivalent intuitive picture of the mechanism for emission from black holes. One can regard the member of the virtual pair that fell into the black hole (say, the antiparticle) as a particle traveling backward in time out of the hole. When it gets to the point at which the virtual particle/antiparticle pair appeared together, it is scattered by the gravitational field into a particle traveling forward in time and escaping from the black hole. If, instead, it were the particle member of the virtual pair that fell into the hole, one could regard it as an antiparticle traveling back in time and coming out of the black hole. Thus the radiation by black holes shows that quantum theory allows travel back in time on a microscopic scale and that such time travel can produce observable effects" (Ibid).

Isn't it all clear and intuitive? There are virtual and unobserved angels and anti-angels. They usually fight and annihilate each other. But anti-angels may fall out of mercy into the infinite black nothing. The angels will be free from the fight with darkness and may escape from it thus bringing "Hawking radiation" to us. There is another possibility that is also quite intuitive. Virtual angels can do whatever they want including time travel. But the paths of these angels are so complicated that it is impossible to keep track of the description. Anyway, the message to take home is that quantum theory allows for everything so it explains all observable effects even the ones that cannot be observed. How can we observe angels that are virtual and unobservable by the definition given by Hawking himself? We should not try to find logic in a model that is trying to combine two models without any physical meaning. It is a belief system and believers do not need any logic and physical meaning.

Unfortunately for physics as science and fortunately for believers in GTR and QM, both speak of entities that cannot be observed: space-time fabric with its black holes that no one can detect by definition or virtual particles that even a particle detector cannot catch also by definition. Hawking changed his belief system and traded black holes for not-so-black ones, but it is a matter of personal choice. The problem is that all these fairy tales about spirits in the void are called physical theories. They did not help Hawking in getting rid of cognitive dissonance. Moreover, he just changed one inconsistent model with infinities for the other with the same problem: "The trouble is … that even "empty" space is filled with pairs of virtual particles and antiparticles. These pairs would have an infinite amount of energy and, therefore, by Einstein's famous equation $E = mc^2$, they would have an infinite amount of mass. Their gravitational attraction would thus curve up the universe to infinitely small size. Rather similar, seemingly absurd infinities occur in the other partial theories, but in all these cases the infinities can be canceled out by a process called renormalization. This involves canceling the infinities by introducing other infinities … It means that the actual values of the masses and the strengths of the forces cannot be predicted from the theory, but have to be chosen to fit the observations" (Ibid).

We have dealt with the quantum inconsistencies and math tricks that put them under the rug in the previous part of the study. But there is no way to escape them, as they compete with GTR in its native domain of gravitational phenomena. QM has its own version of this part of fundamental interactions and is a "creative muddle" of no less scale. According to the quantum version, gravity is caused by virtual particles called gravitons. There is no surprise here as it is a typical quantum magical show that involves virtual spirits in the void. Graviton is considered a massless particle as a carrier of gravitational interaction and quantum of the gravitational field without electric and other charges. Can we discover it? No way. Why? The answer is contained in the definition: particle without mass and charge. Gravity is everywhere, but graviton cannot be detected by the definition given by its creators. Here is an example of a farce: "Unambiguous detection of individual gravitons, though not prohibited by any fundamental law, is impossible with any physically reasonable detector" (Wikipedia, "Graviton"). Nothing forbids us, except for the impossibility.

They even explain why it is impossible: the formation of gravitons will become noticeable only at the energies of the interaction of colliding particles of the Planck energy order. What is the order of this energy? Do not be surprised if they tell you that this order cannot be measured. "In particle physics and physical cosmology, the Planck scale is an energy scale around 1.22×10^{19} GeV ... At this scale, present descriptions and theories of sub-atomic particle interactions in terms of quantum field theory break down and become inadequate" (Wikipedia, "Planck units"). It sounds like theories are inadequate only for this interaction scale, but otherwise, they are adequate. It is a disputable statement, to say the least.

Another article puts it differently: "Most theories containing gravitons suffer from severe problems. Attempts to extend the Standard Model or other quantum field theories by adding gravitons run into serious theoretical difficulties at energies close to or above the Planck scale. This is because of infinities arising due to quantum effects ... Since classical general relativity and quantum mechanics seem to be incompatible at such energies, from a theoretical point of view, this situation is not tenable. One possible solution is to replace particles with strings. String theories are quantum theories of gravity in the sense that they reduce to classical general relativity plus field theory at low energies, but are fully quantum mechanical, contain a graviton, and are thought to be mathematically consistent" (Wikipedia, "Graviton").

When we try to measure phantom entities, they go into infinity, which is a severe theoretical difficulty. To be precise, it is an impasse, as physical measurement becomes an impossible task by definition of the entity. But what does QM suggest? No surprise: another phantom entity "string." What is it? It is another virtual beast from the quantum zoo: a one-dimensional object with length but not width or depth. We will not go deep into the controversy around this idea. Let's just see what this seemingly simple solution leads to and if it is really mathematically consistent.

"Unlike in quantum field theory, string theory does not have a full non-perturbative definition, so many of the theoretical questions that physicists would like to answer remain out of reach ... To construct models of particle physics based on string theory, physicists typically begin by specifying a shape for the extra dimensions of space-time. Each of these different shapes corresponds to a different possible universe, or "vacuum state," with a different collection of particles and forces. String theory as it is currently understood has an enormous number of vacuum states, typically estimated to be around 10^{500} ... In his book Not Even Wrong, Peter Woit has argued that the large number of different physical scenarios renders string theory vacuous as a framework for constructing models of particle physics ... According to Woit, "speculative scientific ideas fail not just when they make incorrect predictions, but also when they turn out to be vacuous and incapable of predicting anything"... In an interview from 1987, Nobel laureate David Gross made the following controversial comments about the reasons for the popularity of string theory: The most important [reason] is that there are no other good ideas around ... In fact, the first reaction of most people is that the theory is extremely ugly and unpleasant ... So I think the real reason why people have got attracted by it is because there is no other game in town" (Wikipedia, "String theory").

The only reason why some people like ugly and unpleasant things is because they do not see any nice and pleasant things around. That is why some theoretical physicists play ugly and unpleasant games. David Gross must know what he is talking about. He is a string theorist and 2004 Nobel prize winner for the asymptotic freedom calculation which is another ugly game around renormalization to hide inconsistencies in quantum field theory (QFT). According to the predictions of the model, interactions should become infinitely strong at short distances. The math hocus-pocus which Feynman as inventor called shell game and a dippy process was used again to hide infinities under the rug. After all, if the inventor got the Nobel prize for this fraud, why shouldn't others follow him? It is a track that proved beneficial for the ugly game leaders. Was it beneficial for physics as a science? They say that it made QFT asymptotically free meaning that calculations did not lead to unreal values. But what were the calculations about? It is the same old story about virtual particles called Quantum Chromodynamics (QCD) that describes the interaction between unobservable quarks. Thus, calculations are about phantoms born from math tricks. Mathematical phantoms produce more phantoms. But they do not bring any benefit for physics as the science about the material world.

We have dealt with these beasts of the particle zoo in the previous part of the study. Here we are interested in other beasts. The quantum gravitation theory says that gravitons are not easy to catch, but their detection is not prohibited. It was a surprising twist for the model that usually dealt with intangible entities that could not be detected by definition. This put the theory under the test. It is not a position to which the believers are accustomed. They had to do something. The model needed a ghost that could not be caught in principle. One-dimensional strings from 10^{100} dimensions of an infinite number of parallel worlds or to be precise from

10^{500} voids came to rescue the model from an inconvenient position of being even theoretically testable and refutable.

One of the founders of the string theory, Lee Smolin, worked on it for decades but eventually realized that it had nothing to do with physics as science and wrote a book with a title that speaks for itself: "The Trouble with Physics. The Rise of String Theory; the Fall of a Science, and What Comes Next" (Smolin, 2006).

Here is how he described this fall: "In quantum theory, there are uncontrollable fluctuations in the values of every quantum variable. An infinite number of variables, fluctuating uncontrollably, can lead to equations that get out of hand and predict infinite numbers … The standard model has a big problem: it has a long list of adjustable constants … If you think of the standard model as a calculator, then the constants will be dials that can be set to whatever positions you like each time the program is run. There are about twenty such constants, and the fact that there are that many freely specifiable constants in what is supposed to be a fundamental theory is a tremendous embarrassment. Each one represents some basic fact of which we are ignorant: namely, the physical reason or mechanism responsible for setting the constant to its observed value … Still, with so many dials to adjust, the theory is difficult for experimentalists to prove or disprove … Not to worry. There turn out to be many different ways to tune the dials to ensure that all the particles we don't see are so heavy that they're as yet unseeable … A regular Noah's ark of particles. Sooner or later, tangled in the web of new snames and naminos, you begin to feel like Sbozo the clown. Or Bozo the clownino. Or swhatever … It is the kind of theorizing that can't fail, because any disagreement with present data can be eliminated by tweaking some constants … No more reliance on experiment to check our theories. That was the stuff of Galileo. Mathematics now sufficed to explore the laws of nature … The constants that could be freely varied in the standard model were translated into geometries that could be freely varied in string theory. Nothing was constrained or reduced. And because there were a huge number of choices for the geometry of the extra dimensions, the number of free constants went up, not down … But the exact combinations seen in nature did not come out of the equations. From here it got worse … There were lots of instabilities, which manifested themselves in lots of extra forces and particles … All we had was a list of hundreds of thousands of distinct theories, each with many free constants … Worst of all, there was not a single prediction made that might be confirmed or falsified by a doable experiment … When I complained about this to some of the leaders of string theory in the mid 1990s, I was told not to worry, it was just that the theory was smarter than we were. We cannot, I was told, ask the theory questions directly and expect answers. Any direct attempt to solve the big problems was bound to fail … But, simply put, once you reason like this, you lose the ability to subject your theory to the kind of test that the history of science shows over and over again is required to winnow correct theories from beautiful but wrong ones. To do this, a theory must make specific and precise predictions that can either be confirmed or refuted. If there is a high risk of disconfirmation, then confirmation counts for a lot. If there is no risk of either, then there is no way to continue to do science … The stakes are to accept the landscape and the dilution in the scientific method it implies or give up science altogether and accept intelligent design as the explanation ... This is painful for many who have invested years and even decades of their working lives in string theory" (Smolin, 2006).

It is no surprise that theoretical physicists who chase the phantoms of their models at some point begin to feel themselves as clowns. There are two ways to get rid of this uneasy feeling: look for the physical reason and mechanism responsible for the observed or call for a divine intelligent Being responsible for everything. Newton made the second choice when he struggled with the puzzle of gravitation. Many modern theoreticians follow his suit but still call themselves physicists. Even the founding fathers of the quantum Wonderland were amazed at the achievements of their students. Richard Feynman wrote to his wife about one of the conferences on quantum gravity: "Because there are no experiments, ... the result is that there are hosts of dopes here ... and it is not good for my blood pressure. Remind me not to come to any more gravity conferences!" (Feynman, 1988).

While theoretical conferences on gravity were full of dopes talking about delusional constructs virtual entities that cannot be tested with any experiment, astronomers were busy with "the stuff of Galileo," i.e., testing reality. Observations of the motion inside the galaxies led to a striking result: the orbital speed of stars at the periphery of galaxies is close to the speed at the center. This is fundamentally different from the observed planetary motion in the solar system, where speed decreases with distance from the center. By the 1980s, it was finally clear that galaxies did not decrease orbital velocity with increasing distance from the center of mass. In none of the observed galaxies, rotation curves corresponded to predictions of standard gravitational models. According to them with such parameters of motion, the galaxies should be falling apart, but they keep on rotating and preserve their structure.

Two explanations are possible: there is some invisible mass that is responsible for such interactions or mainstream theoretical models are inadequate. The mainstream chose the first one as it could not accept the truth about the model. In the 1930s, the astronomer Fritz Zwicky hypothesized that some hidden mass affects visible objects and distorts their trajectories. Calculations within the existing models showed that its mass exceeds the mass of the entire visible part of the Universe. At first, the idea was so contrary to the "good old" ideas about gravity that the mainstream just chose to ignore it. A few decades later, the data flow increased so much that it became impossible to deny the discrepancy between Newton's law and observed phenomena. Zwicky's hypothesis grew into the theory of dark matter.

The Standard Cosmological Model acquired a new name: "The ΛCDM (Lambda cold dark matter) or Lambda-CDM model is a parameterization of the Big Bang cosmological model in which the universe contains three major components: first, a cosmological constant denoted by Lambda associated with dark energy; second, the postulated cold dark matter (abbreviated CDM); and third, ordinary matter" (Wikipedia, "Lambda-CDM model").

We have seen that lambda was needed for adjusting the model to any observable energy phenomena in the Universe and was a typical math trick to save the inadequate model. Anything that did not correspond to the predictions of the model was thrown into a trash bin of "dark energy." The name reflected the darkness of the model's explanatory gap. The same happened with "dark matter." All phenomena of interactions that the model could not account for were simply called something dark and unknown. This is the actual state of the Standard Cosmological Model based upon the "most beautiful of all existing physical theories," the holy scripture of the GTR.

Lee Smolin commented: "Things have become even more mysterious. We have recently discovered that when we make observations at still larger scales, corresponding to billions of light-years, the equations of general relativity are not satisfied even when the dark matter is added in ... Again, there are two possible explanations. General relativity could simply be wrong ... Or there is a new form of matter or energy ... This strange new energy, which we have postulated to fit the data, is called the dark energy ... Fully 70 percent of the matter density appears to be in the form of dark energy. Twenty-six percent is dark matter. Only 4 percent is ordinary matter. So less than 1 part in 20 is made out of matter we have observed experimentally or described in the standard model of particle physics. Of the other 96 percent, apart from the properties just mentioned, we know absolutely nothing ... There is now a standard model of cosmology, just as there is a standard model of elementary-particle physics. Just like its counterpart, the standard model of cosmology has a list of freely specifiable constants — in this case, about fifteen. These denote, among other things, the density of different kinds of matter and energy and the expansion rate. No one knows anything about why these constants have the values they do. As in particle physics, the values of the constants are taken from observations but are not yet explained by any theory ...Many people have thought hard about this for years, and we are more or less nowhere ... If there is no way to account for a phenomenon on the basis of what we know, then maybe this is a sign that we need to look for something new" (Smolin, 2006).

It all boils down to a childish horror story: on a dark day, something dark was sitting in a dark place. Or as one popular science TV series concluded: "As the world of light was formed by dark matter, perhaps the whole Universe was formed by an even darker entity. Perhaps dark matter has an even darker shadow. Maybe behind dark matter, there is an even darker entity that rules the world" (Through the wormhole, Discovery Science channel). Perhaps it is not about some dark forces out there but about our model of the gravitational interaction that is in the dark? The need to replace inadequate models is obvious to many theorists and, of course, to practical researchers who really feel themselves in the dark without any model to rely on.

But what do theorists suggest? For example, Smolin, who criticized the mainstream models, is one of the founders of the Loop Quantum Gravity (LQG) theory. Does it offer something new? Not really. It just tries to incorporate Einstein's geometric formalism into quantum mechanics. Within this model, gravity is the result of the special structure of space-time fabric composed of loops woven into "spin networks" or "spin foam." The phantom began to loop instead of curving.

Smolin explained: "To see how beautiful and simple this is, note how these photon-like and graviton-like particles arise from strings. Strings can be both closed and open. A closed string is a loop. An open string is a line; it has ends … Thus, you get particles and forces alike from the open strings, and if the theory is designed cleverly enough, it can produce all the forces and all the particles of the standard model. If there are only open strings, there is no graviton, so it seems as though gravity is left out. But it turns out that you must include the closed strings. The reason is that nature produces collisions between particles and antiparticles. They annihilate, creating a photon. From the string point of view, this is described by the two ends of the string coming together and joining. The ends go away and you're left with a closed loop. In fact, the particle-antiparticle annihilation and the closing of the string is necessary, if the theory is to be consistent with relativity, meaning the theory is required to have both open and closed strings. But this means it must include gravity. And the difference between gravity and the other forces is naturally explained, in terms of the difference between open and closed strings" (Smolin, 2006).

Here is another "beautiful and simple" model. It all comes down to some one-dimensional entities called strings being closed or open and thus producing all particles and forces, but only if the model is clever enough. Was it clever enough? It was too clever. Instead of simplifying things and getting rid of hordes of virtual entities, arbitrarily set constants and infinities crawling up in calculations, things got worse as the model tried to use the same math tricks. It is the same "extremely ugly and unpleasant game" that many theoretical physicists play and make the string theory rise meaning the fall of science. What comes next?

Chapter 5

Solving the Puzzle

Solved puzzle becomes trivial.

Lev Landau

We can describe the puzzle of gravity as a list of paradoxes:
1. One-way paradox.
It seems that this fundamental interaction has only one side — the attraction. The apples fall on our heads and do not fly away into the sky. We are used to calling it gravity from Latin gravitas — weight; downward movement, attraction. We interpolate the local one-way effect on the non-local cosmic bodies and think of them as gravitating to each other. Classical gravity law has only a positive solution. But this contradicts the reality that speaks of things being in various states of interaction: attraction, repulsion, and balance. If only one-way gravitation worked, everything would collapse into a gravitational singularity. To be precise, the Universe would not exist in the first place.

When we discover that this interaction has a repulsive direction, we tend to go to the other extreme and talk about a Universe that is expanding into nothing and nowhere. Thus, we ignore the obvious facts of balanced states of things in our native solar system and the observed far-away galaxies. The reason for our inability to combine all sides of interaction in one model is the lack of understanding of the mechanism that can provide for them.

Here is a typical description: "Gravitation is the mutual attraction between all masses in the universe, also known as gravitational attraction. Gravity is the gravitational attraction at the surface of a planet or other celestial body" (Wikipedia "Gravity"). It speaks only about attraction. The fact that comparatively small bodies are attracted to the surface of a big celestial body does not mean that all things in the Universe only attract. Once again: we would not be witnessing the existing order if only one side worked. There would be no order and no one to witness it. There would be one black hole or, to be precise, one nothing. Lucky for us, black holes (local or universal) do not exist. They are the phantoms produced by our inadequate models of the world that stemmed from our ignorance about how the "device" actually works. If we understand the mechanism the paradox will evaporate. We would be better off if we called this level of energy interactions by another name. But the name stuck so TEH uses it for historical reasons. But we should keep in mind that this interaction is no different from other levels and has three sides to it.

2. The external force paradox. It seems that bodies experience an external force acting upon them so that they are attracted to each other. We are used to calling it the force of attraction. It is another error that stemmed from our ignorance about the mechanism. The concept of force in physics is just a convention that makes the observed interactions with the push-pull effect open for mathematical description. We can estimate its magnitude and direction, which makes the concept useful from a practical point of view. It is also convenient for making analogies between different manifestations of the same effect. That is why we combine them with one word "force" and just apply different names (gravitational, electrical, magnetic, weak nuclear, strong nuclear, normal, frictional, applied, etc.). But the concept of force leads to a paradox: it seems that something acts upon things and it is here, there, and everywhere but cannot be found. Thus, the concept of force becomes not a physical but a metaphysical one.

This has been appreciated a long time ago and modern physics uses the concept of interaction. Things interact and demonstrate push-pull effects or a balanced state. Unfortunately, this correct change in perspective has not yet led to the solution of the paradox. The reason is the same: lack of understanding of a mechanism that is behind all observed effects. As a result, physics got back to the concepts of transcendental entities acting upon things to make them interact. It does not matter that they got new names (virtual particles, space-time fabric). The essence remains: they are here, there, and everywhere but cannot be found. We are back to metaphysics and we should either believe in these entities or not believe in them. This is the fall of science. To make it rise we should solve the external force paradox. This concerns not only gravitational interaction but all known types of interaction. For this, we need to uncover the mechanism that is inherent to things and not transcendent to them.

3. The weakness paradox. Gravitation is called the weakest of all fundamental interactions and is evaluated to be 10^{38} times weaker than strong nuclear interaction, 10^{29} times weaker than weak nuclear, and 10^{36} times weaker than electromagnetic one. Thus, they say, that it has no significant influence at the microcosmic levels but exerts a big influence at the macrocosmic levels where masses are huge. How can such a weak force be so strong that it moves planets, stars, and galaxies? We can put it differently if we drop the word "force." How can interaction be weak and strong at the same time? Classical physics says that it is because the parameters in the gravitational law are mass and distance. It is, of course, a fact of life: the more massive the object the stronger the interaction and the further the object the weaker the interaction. But this is not an explanation, but

a description of the observed effects. To solve the paradox, we need to uncover the mechanism that will explain why mass (energy) and distance influence the magnitude of the interaction.

4. The omnipresence paradox. It seems that gravity has infinite range and speed but no medium to propagate through. This leads to an obvious contradiction that bothered classical physicists. As we remember, Newton could not solve this problem and turned to God as the ultimate explanation. When a physicist explains physical things with non-physical concepts cognitive dissonance is inevitable. It bothered all classical physicists. Then came Einstein who tried to overcome this dissonance by inventing an omnipresent space-time fabric that influences the bodies and makes them move as they do. But this produced more paradoxes instead of solving the basic one.

STR spoke of light as the movement of massless particles in the void but GTR spoke of gravitation as something happening with massive bodies in an environment. It does not matter that Einstein did not uncover how the "device" actually worked and just described its geometry. His space-time fabric is essentially an environment. The paradox is that the emptiness of Einstein's model of light and the fabric of Einstein's model of gravity occupy the same place. Is the space without visible objects an emptiness that cannot be characterized by any physical quantities or is it some fabric that has a physical nature to make it an environment for physical interactions? As a result, he had to introduce a special kind of medium that was not physical and could be everywhere, even in the void, but still be the intermediary for the interacting physical bodies. Physics again turned into metaphysics. The cognitive dissonance was not overcome.

There was another paradox. Within STR the speed of light is an ultimate absolute, and Einstein was very specific that nothing can move faster than light. But GTR speaks of space-time fabric that implies instantaneous action at a distance in the form of a distortion of the geometry of this fabric creating a gravitational effect. Einstein was critical of any hints of non-locality and instantaneous action within other models but his model violated his own principles. To this day practical physicists try to measure the speed of gravity and estimates differ. Some say that it is the same as the speed of light. But studies of binary stellar systems have shown that gravity travels several orders of magnitude faster than light (Hotson, 2002).

We have dealt with the issue of the speed of light earlier. What we're talking about here is that a paradox arises: the same manifestations that are seen in all interactions seem to come from different mechanisms that are completely incompatible with each other. Einstein was bothered by this contradiction between his models of different fundamental interactions to the end of his life (more on this later). Moreover, his model is incompatible with other models of fundamental interactions. This contradiction bothers physicists to this day. They try to reconcile GTR with SM but fail or create an illusion of success by inventing more phantom entities. The universal physical mechanism remains an enigma.

The omnipresence paradox concerns all interactions as they seem to happen everywhere without any intermediary. Even though Einstein abolished the notion of a medium filling all the space (his first step in STR), physicists this way or another spoke of a medium. The name "ether" became a taboo but the prophet did not forbid the word "field" so they used it extensively in all models. Moreover, when Einstein realized all the contradictions, he started to work on a unified field theory himself. He failed and there were reasons for that (more on this later). Even if we call an all-encompassing medium with a politically neutral name "field," this does not relieve us of the responsibility to explain the mechanism that operates in this field.

If we apply the term "field" we can call TEH the unified field theory as it speaks of a universal mechanism for all interactions. But whatever name we use for an all-encompassing medium it is still about some intermediary for any interaction at a distance. But the fact is that all interactions occur at some distance. Nuclear interactions have their own, though microcosmic, distances. So, we are again facing a paradox: interactions happen everywhere but they do not have a medium for propagation. Or do they?

It looks like we should untangle all the knots starting with the last paradox. A logical chain is evident: if energy interactions are ubiquitous and happen at a distance, they are mediated by an all-encompassing energy environment. If we ignore this medium, we face all the paradoxes and any attempt to think of a mechanism of interaction will lead to the creation of phantoms that act in the void. We have created so many phantoms that not a single theoretical physicist can enumerate them. Noah's ark is overcrowded with phantom beasts. We don't need hundreds of thousands of theories to describe one-dimensional looping strings, nor even a single theory to describe hundreds of virtual particles, or to describe one fabric of curving geometric illusion.

We can paraphrase Einstein. Here is the quote once again: "The first step we must take is to *give up the ether* … The only way to arrive at a satisfactory theory is to give up the notion of a medium filling all the space" (Einstein, 1910). It was a step in the abyss and the fall of science. We have to get out of this abyss and rise the banner of science. The first step we must take is to *give up the void*. The only way to arrive at a satisfactory theory is to give up the notion of an empty space filling all the space.

It sounds easy from the physical meaning perspective but not so easy from the political perspective. But if we want to see the rise of theoretical physics we should not bother with politics in the "scientific community" that has turned into a religious corporation. The second step to take is to reveal the mechanism that works in the medium filling all the space.

We have taken both steps in the previous volume and will continue along this road here. We have shown that the same mechanism of frequency-phase coupling (synchronization) of energy oscillations is behind the planetary, atomic and subatomic levels of interaction. Thus, it can be called a universal mechanism of fundamental interactions.

We can paraphrase Blandford. Here is the quote again: "The theory of general relativity (including λ) provides a singular prescription for describing classical gravity and nongravitational physics in this environment. There seems to be no need for baroque variations and empirical corrections over its domain of applicability. It has grown from an essentially intuitive quest to providing the foundation for high-energy astrophysics and cosmology, in which it is routinely considered to be correct" (Blandford, 2015).

Theory of Energy Harmony provides a description and explanation of a mechanism that works for classical gravity and nongravitational physics in this environment (not in the void). It does not include any arbitrary variables and has self-consistent simple math solutions that describe harmonious ratios of frequency-phase coupling interactions that manifest themselves in attraction, repulsion, and balanced states of interacting elements. There is always a need for variations and empirical corrections as it is not a belief system but a scientific model that cannot be final for all time and should be tested against reality. It is intuitive and based on simple analogies that unite well-known and local phenomena with distant and less studied ones. It provides a foundation for cosmology that should not routinely consider it correct but test its correctness. The facts we know about the solar system and some galactic phenomena prove that the mechanism offered by TEH is both local and non-local. The prediction is that the more we know about astrophysics the more confirmations we will get that the same mechanism is behind the observed. Thus, TEH provides the explanatory and predictive base for macro-, meso- and microlevels of the world.

For example, new findings by astronomers showed that there was a phenomenon of counter-rotation in galaxies. Some gas and stars move in the opposite direction to the rotation of the rest of the galaxy. In addition, some galaxies rotate clockwise and some counter-clockwise. Neither the classical model nor the new cosmological model based on the "singular prescription" of GTR predicted such dynamics. But such positive and negative helicity can be observed even in a bathroom. We are back to the hydrodynamic analogy.

Contrary to popular belief that the direction of the water flow vortex in the sink depends on location in different hemispheres, you can get different rotations of the vortex in sinks located in the same hemisphere. It all depends on the experiment's parameters and conditions: the configuration of the system, surface structure, temperature, state of the liquid, air flows, etc. In other words, it is a dynamic phenomenon with many degrees of freedom.

We can assume that the phase portraits of galaxies, as ensembles of oscillators, result from the interaction of these oscillators. The interaction, which we so far habitually call gravity, although it has an attraction vector and repulsion, depends on the oscillators' parameters and the medium. In this case, an explanation of the observed dynamics will not require mysterious dark energy or phantom fabric involvement. Different parameters can influence the swirling of vortices in the bathroom and galaxies. But if they have common patterns of external manifestations, then a common mechanism is likely.

By the way, the vortex motion as a whole has not yet received a worthy model either in hydrodynamics, or aerodynamics, or gravitational dynamics. Sometimes the movement of the vortex in the sink is explained by the Coriolis effect. It is about a body moving within a rotating system and shifting in its trajectory in a direction perpendicular to its movement. For example, the rotation of the Earth shifts the Foucault pendulum, and it draws trajectories displaced from straight lines. This way, the Foucault pendulum confirms the daily rotation of the Earth.

The same effect is also manifested in the displacement of cyclones in the atmosphere, which in the Northern Hemisphere rotate counter-clockwise and in the Southern Hemisphere clockwise. The ease of analogy gave rise to the myth of a similar dependence of the direction of the vortex in the sink on the hemisphere. But there is one problem: the angular velocity of the Earth's rotation is so slow (revolution per day) that the effect has a strong manifestation only at large distances and an extended period. It is maximum at the poles and minimum at the equator for the same reason: the angular velocity of rotation around the axis is different. It affects cyclones and currents in the ocean. The Foucault pendulum is not as large as a cyclone, but its displacement is very slow. In the original experiment at the latitude of Paris, the pendulum's oscillation plane rotated by 11°, i.e., made a full revolution in 32 hours.

So, the reality is this: there are interacting oscillatory phenomena with different parameters, the combination of these parameters is key for the manifestations of such an interaction. This simple fact seems to disappear from sight, and the movement in a particular direction is explained solely by the Coriolis effect with respect to the rotation of the Earth. This "explanation" can be found even in textbooks.

One simple question: what is the rotation speed of water in the sink? Even if you did not measure it, there is no doubt that it is much faster than the displacement rate of the Foucault pendulum trajectory. The Foucault pendulum size is not accidental: it is suspended under the cathedrals' high ceilings because only in this way it is possible to catch the minimum displacements. The funnels and pipes in the bathroom are usually much more modest in scale, but the rate of displacement is obvious even to the naked eye. The conclusion is simple: the Coriolis effect has nothing to do with it. If it plays its role, this role is negligibly small (thousands of times) compared to other factors in such a complex, albeit seemingly simple, water circulation system in our sinks.

Galaxies have many vortexes. We observe incremental accretion rotating disks with polar jets (energy jets along the axis of the disk rotation) everywhere in stellar space (galactic nuclei, the birth of stars, old neutron stars, white dwarfs, protoplanetary structures and other objects). The rotation around the axis and the pulsation of the jets along the axis have their own frequency, indicating the source's oscillatory nature. Observations show that accretion disks have quasi-periodic oscillations. The period of these oscillations has a scale inversely proportional to the mass of the central object. Overtones of different objects have different frequency ratios. It is a description of the presence of frequency-phase relationships of oscillatory processes. Stars play their music for us, but we ignore it. The same goes for music of the earthly matters. We observe tornadoes on Earth with similar manifestations, but they remain a mystery to us.

If you think that the vortex in the sink has been described by physics long time ago, you are mistaken. Physics is so preoccupied with its virtual entities that it is not up to ordinary everyday oscillations, be it galaxies or water in the sink. For example, Richard Feynman, in his famous textbook "The Feynman Lectures on Physics," assigned a modest role to hydrodynamics in general. Not because everything is clear and not interesting there, but rather because it remains "uncharted land" (more precisely, water). Having devoted a lot of space to electrodynamics, the author gives a brief overview of hydrodynamics and notes: "The hydrodynamic equations are often closely analogous to the electrodynamic equations; that's why we studied electrodynamics first. Some people argue the other way; they think that one should study hydrodynamics first so that it will be easier to understand electricity afterwards. But electrodynamics is really much easier than hydrodynamics" (Feynman et al., 1964-1966).

Describing how fluid oscillates in a vertical tank in the presence of a drain hole below, Feynman claims that you must first spin it with a stick to give it angular velocity. If this external driving oscillator (stick) stops, then the driven oscillation (vortex), according to the author's description, should stop due to the viscosity of the liquid. The flow will cease to be a vortex. This follows from the equations of theoretical physics. But how does water flow in real life? No stick is needed. The flow becomes a vortex by itself, remains such, does not slow down, and even accelerates. No "prime mover" is required because there are self-oscillating processes in which there are many more factors besides the viscosity of the medium. One has only to open his eyes and look into his bathroom sink to understand the inadequacy of theories and formulas that Feynman is using.

This textbook is considered to be "perhaps the most popular physics book ever written" (Wikipedia, "The Feynman Lectures on Physics"). It is a peak of the evolution of mathematical physics of the twentieth century: loss of the reality testing function and the collapse of the model's coherence (in clinical terms, these are symptoms of a psychosis). Feynman himself made the diagnosis: "Today our theories of physics, the laws of physics, are a multitude of different parts and pieces that do not fit together very well. We do not have one structure from which all is deduced" (Feynman, 1965).

But if there are general patterns of manifestations of interactions, then perhaps there is a common mechanism for these interactions? Does it make sense to divide these interactions into fundamental and not so fundamental, into different forces and fields? We split everything into different parts and pieces and wonder why they do not fit together very well. But the process of cognition includes analysis and synthesis. There is a time to cast away stones, and a time to gather stones together (Ecclesiastes 3:5). Let us try to return to the basic formulas that describe the phenomena of gravitational and electromagnetic interactions and understand where there was the source of the misunderstanding that first put these phenomena into different fields and then became a stumbling block for attempts to unite them into a single theory.

Newton's law:

$$F = G \cdot \frac{m_1 \cdot m_2}{R^2}$$

Coulomb's law:

$$F = K \frac{|q_1 \cdot q_2|}{R^2}$$

Drawing up a formula describing an interaction's manifestations does not mean understanding the physical essence of this interaction. Newton admitted that he did not know the nature of gravity and did not invent the spirits responsible for it (except for one "intelligent Agent," invented a long time ago). He gave an equation and left the question of the mechanism to future generations. Coulomb did the same for electrical phenomena.

If mass and charge are discrete parameters by which we measure the continuous energy processes, then there is no problem in designating gravitational mass and electric charge in the same formulas as different symbols. But one should not lose the general physical meaning of the observed phenomena at different energy levels. Suppose we accept the working hypotheses that mass and charge express the level of energy oscillations and interactions. In that case, we begin to understand why the gravitational force can be the repulsive force at the macro level, just like the attraction force and repulsive force may be the manifestation of interaction at the micro-level.

Factors influencing the interaction are the characteristics of the oscillators and their phase portraits, parameters of the propagation medium and distance. This sounds as an obvious statement, but one of these defining parameters is excluded from the formulas: mediating energy environment. It is stubbornly ignored, and "proportionality coefficients" are inserted instead: the gravitational constant and the electric constant. They create the effect of the formula correspondence to the observed phenomena. It looks like there are no factors except mass or charge and distance.

An arbitrary variable called "constant" becomes a substitute for describing a multifactorial process. No wonder formulas work only for some instances when we can reduce the number of variables. At the same time, they create the illusion of explanation: if the equation has a solution, then we understand the physics of the process. But the removal of the environment and the introduction of the void leads to the need to create virtual entities flying in this void. The power of such faith is so great that it can suppress common sense and ignore facts.

However, the facts indicate that all manifestations of interaction, including attraction and repulsion, arise as material phenomena in a real energy environment. There are no special "forces" (gravity, electromagnetism, etc.), but there are manifestations of the same process and different energy levels at which it occurs. This idea is necessary to create the basis for a unified theory of interactions. But it is not sufficient. There remains a "trifle": we have to look "under the bonnet" and understand the mechanism of such interaction.

Coulomb's law has a charge modulus because it describes both attraction and repulsion. Although it has an identical form because of the same regularities, Newton's law ignores repulsion and speaks of gravitation only. Thus, the model contradicts the observed phenomena: cosmic bodies attract and repel and establish stable systems with an attraction/repulsion balance. While trying to answer the inevitable question of why doesn't everything fall together in a mess, it has to introduce new forces and even come up with antimatter, negative gravitational mass and other auxiliary variables. Newton's formula can lead to this solution:

$$F = -G\frac{m_p m_a}{r^2}$$

Objects with negative mass will, like "point charges in the void," be repelled by positive masses and attracted by negative ones. Mathematically, there is no problem that the mass can be negative, but physically this is absurd from whatever perspective we look at the notion of mass. The inert mass is a measure of the body's inertia, i.e., the relationship between the reaction of the body and the acting force. Newton's second law expressed this quite clearly: the force applied to the body is equal to the acceleration multiplied by the mass $F = m_i \vec{a}$. The formula spoke of linear interactions and did not consider various factors that adjust direct proportionality. In this form, the absurdity of an attempt to introduce a negative mass becomes obvious: a resting inert body should accelerate in the opposite direction to the applied force. If the body has its own momentum, its vector is opposite to the momentum of the applied force and amplitude is greater, it will move in the opposite direction. But then, this is no longer an inert mass but a completely different physical category. If we apply the negative mass concept, we either make a category error or create phantoms that contradict empirical evidence. In both cases, our model breaks away from reality.

Suppose we accept the hypothesis that gravity is a real energy interaction in a real medium and not a magic force acting in the void at any distance. In that case, the influence of the medium on the propagation of such interaction can be twofold: it provides propagation and speed, but affects the trajectory and limits the range of propagation and interaction of these specific sources and receivers of these waves. This makes the gravitational interaction, like any other at different energy levels, limited and relative to the characteristics of the interacting objects themselves and the environment.

Given that all objects are both sources and receivers, the resulting form of interaction is determined by wave processes' interference pattern. Since there are many elements involved in the process, the picture can be very complicated. But complexity does not mean randomness. If we are talking about wave interactions, a stable, coherent synchronized structure remains balanced and dynamic. For example, the solar system is such a coherent structure with harmonious combinations of frequencies and phases, allowing the system's objects not to form separate pairs but to remain in the general ensemble of dance and music of the spheres. Each element's contribution is determined by its energy, frequency, and phase characteristics, which constitute what we are used to calling mass. The common phase portrait is determined by the characteristics of the environment and its "quantity" between interacting objects (we call it distance). Therefore, the manifested effect of such a wave interaction is the gravitational force's dependence on mass and distance, described by Newton in the form of the Law of universal gravitation.

However, no special gravitational field or space-time fabric is needed for such an interaction since this interaction is only one of the levels of energy oscillatory processes. One does not have to use the concept of "anti-gravity" or "fabric curvature" to explain why, at certain interaction parameters, the oscillators do not fall on each other but form stable synchronized systems, preserving the individuality of the phase portrait of each element. But with different parameters, they can also merge into dense structures or scatter.

Einstein wrote about the force of gravity as an illusion: nothing is attracted, but only "floats" in the geodesic lines of the space-time fabric. He was right: indeed, gravity can be called an illusion, but in the sense that there is no separate "force" of attraction, which has its special laws, fields and carriers. Gravity and other forces invented by physics theories as a way to describe phenomena are phantoms of inadequate modeling. They are illusory, but not because there are no phenomena of attraction/repulsion, but because theories create entities responsible for phenomena, while these are not entities, but processes of interaction of different energy levels and forms of matter. The illusory nature of the GTR and SM concepts becomes an obstacle to unified theory of all interactions. But if we look at gravity, electromagnetism, nuclear and the rest of interactions as manifestations of oscillations at different levels of the same energy environment, then we can build such a unified theory. This is not a theory of everything, but a theory explaining a wide class of similar phenomena. Their similarity is a consequence of the universality of the mechanism.

Why did the apple fall on Newton's head? If the answer is "the force of gravity," it creates a spirit that explains all the manifestations for which it is responsible and itself. The very concept of gravity is tautological: objects gravitate because the force of gravity acts. It is no accident that Newton, who described the manifestations of gravity mathematically, admitted that he did not know the mechanism and was forced to lean towards the old idea about the higher intelligent Agent responsible for everything. If the physical mechanism of gravity is not explained, then inevitably, there is a reference to a single metaphysical "explanation" for everything.

The answer to the question why we are attracted to the Earth in the model proposed here is simple: the Earth is a synchronizer for all objects as energy oscillators in the zone of its influence. It is also in the synch with other oscillators of the solar system. A steady structure is determined by its elements' energy parameters, frequency-phase relationships, environment, and distances. We cannot establish equal relations with the Earth for a simple reason: our energy parameters (mass) are too small. In general, all objects on Earth are so small in relation to it that the illusion of the same gravitational acceleration of about 9.8 m/s^2 is created. This is expressed in the fact that in a rarefied medium (vacuum), where the influence of air resistance is minimized, the fall of bodies of different masses occurs simultaneously. A feather and a lead ball will fall in a vacuum at almost the same time.

This is called the Galilean equivalence principle or the law of the universality of free fall. It states that all objects in a given point of gravitational interaction undergo the same acceleration in a vacuum independent of their properties (composition, structure, rest mass). It is also called the weak equivalence principle as it concerns only local effects for comparatively small size bodies. Tests of this principle are usually done by dropping different objects in the environment where other influences on the falling object are excluded (for example, in Fallturm Bremen tower). Thus, this principle concerns only specific experimental conditions that measure relations between Earth and objects in its zone of influence.

Einstein had his own variation of the equivalence principle. He offered a thought experiment with an elevator. If a person drops an object in a stationary elevator on the Earth it will fall to the floor with a specific acceleration. If a person drops the same object in an elevator that is flying in space with the same acceleration as the free fall on the Earth, the floor will move towards the object and it will seem that the object is falling with equal acceleration. Thus, we cannot tell the difference between being in the gravitational interaction or not. He postulated that this holds true for any physical measurement though his thought experiment was an abstract setup with all the subtleties of reality not taken into account.

The believers in GTR say that the experiment concerns only the "freely falling local laboratory" (the new euphemism for an elevator) which must be small compared to the variations of gravity. But to compare the actual result in such a lab with the free fall on Earth it must have the same mass as the planet. However, it must not be a source of its own gravity, otherwise, the experiment does not tell anything about the equivalence of gravity and acceleration. The experiment implies an impossible condition: the absence of any other interactions in the lab and its closure to the outside world. Einstein also assumed constant acceleration without considering the real mechanics of generating it. As usual, he was not concerned with how the device actually worked but produced postulates for a religion that the followers must believe. In general, the "elevator experiment" is not actually an experimental test and does not prove anything. It is full of internal contradictions and contradicts reality though on the surface it seems to be simple and intuitive. Even the mainstream propaganda has to acknowledge that "the Einstein equivalence principle has been criticized as imprecise because there is no universally accepted way to distinguish gravitational from non-gravitational experiments" (Wikipedia "Equivalence principle").

But Einstein considered the idea that gravity is the same as acceleration the "happiest thought" of his life. It allowed him to combine STR which did not include gravity at all with GTR which excluded gravity as an interaction or force and made it just a geometry of space-time. Now the free-falling objects were just moving along the locally straight paths in curved space-time. They were sliding easily and happily with the same acceleration. The rest was easy: describing the metric of space-time with tensors that do not have real solutions. From here the metaphysics of GTR began.

Let's get back to physics again. We have to explain physical phenomena with a physical interaction mechanism. At first glance, the phenomenon of the same acceleration of objects with different inertial masses seems counterintuitive and even paradoxical. How can a feather and a lead ball fall simultaneously? The truth is that they do not fall at the same time, but almost at the same time.

Within TEH, gravity is not a one-way attraction and is not a one-way force acting from one body to another. It is a "two-way street" where the parameters of all participants and their ratios are important for the result. Thus, the paradox is resolved. The masses of the feather and the ball relative to the mass of the Earth can be taken to be equal as a first approximation. Therefore, measurements at a certain level of accuracy will give the effect of almost the same fall time. But the ball will arrive at the bottom some nanoseconds earlier. If we imagine an experiment with the fall of a feather and a body weighing a quarter of the Earth's mass, then the difference will be noticeable to the naked eye. If an object with comparable mass to that of the Earth were to move towards it, then we could even observe the corresponding movement of the Earth. In any case, all participants have their part in the interaction scenario.

Gravitation is not equivalent to acceleration. Einstein's "happiest thought" was a plain category error. Gravitation (attraction) is one side of the fundamental interaction manifestation. Acceleration is the rate of change of velocity, the dynamics measurement parameter. These are different categories and they cannot be compared to each other. We cannot say if they are equal or not. Einstein's equivalence principle does not solve the riddle of the classical equivalence principle. It confused things even more and produced the phantom of space-time fabric. One logical error led to another one.

The fall of objects is a process that is due to many factors, including gravitational attraction, but it does not mean that their acceleration and attraction are equal. Any thought experiment that makes them equal would be a mix-up of processes and concepts. If we mix things that are not in the same category, we inevitably need to invent a phantom that "glues" them together. Thus, for Einstein, autumn leaves falling to the ground follow some paths in the curved space-time fabric. The elevator experiment sounds even more ironic: "The mystery of gravitation puzzles us, except those who have the luck to fall in an elevator, and even for them knowledge comes too late. They weren't falling at all: just curving" (Leacock, 1956).

The reality is a little bit more complicated. The leaves and elevators fall to the ground for many reasons. Gravitational interaction is one of them and it is not an illusion produced by paths in the phantom fabric. It is a physical phenomenon that is limited by the parameters of the interacting elements and the environment. The distance is one of the leading factors even if the speed of propagation is enormous. When the distance changes, the coupling strength of interacting energy oscillations changes. This explains why the attraction force is different at various points on the Earth's surface. It is the weakest at the equator because this surface is furthest from the center of the Earth meaning the vector of interaction is the longest in this case. The same goes for objects flying away from the ground.

Moreover, the attractive side of this interaction may be overcome. It does not mean that there is anti-gravity as some opposing force. Any interaction has three manifestations: repulsion, attraction and balance. It is all about interacting oscillators establishing certain ratios of their parameters. Let's take a real "elevator" (spaceship) starting into open space and getting away from Earth's attraction. The engineers have to calculate its technical parameters and the most effective flight path. These are real subtleties of a real experiment. The equations include the engine thrust, the spacecraft's speed and mass, the angle of inclination of the speed to the horizon, the equatorial radius, the compression ratio, the coordinates of the launch pad, the angular velocity of the Earth's rotation, acceleration, a set of aerodynamic parameters and much more. The practitioners say: we know that gravity cannot be ignored or turned off, but it can be overcome.

But what is the basis of calculations? If we remove the influence of all other factors, i.e., accept the starting conditions as ideal, and the Earth's attraction as the only obstacle that must be overcome, then the trajectory is determined according to Kepler's laws describing the shape of the orbits (phase portrait) of cosmic bodies, a change in the linear velocity of the path (phase dynamics), calculation of angular velocity and oscillation frequency. All other parameters become additional variables of local conditions for adjustments to the calculations (for example, the state of atmospheric processes). When estimates are made for an interplanetary trajectory with respect to unknown local conditions, the base of Kepler's laws is applied, and correction is made for the alleged factors and parameters of the distant planet. But in no description of the calculation of flight parameters you will find mention of space-time fabric curvature. Einstein's tensors do not have any practical use not just because they do not have any solutions but because there is no space-time fabric.

Mainstream models of gravity are not even wrong. They are just not adequate to reality and irrelevant for practical use. If you ask practicing physicists and engineers how they measure the fabric's curvature during flight preparation and how they modify this fabric to overcome gravity, they will look at you with a strange expression on their faces. But you have another version of the question. According to SM's dogmas, the carriers of gravitational interaction are gravitons, i.e., bodies are attracted to the Earth because they "swallow gravitons" (exchange the gravitational carriers). How do engineers calculate how many gravitons a spaceship has to spit out so that it flies into space? How much it is necessary to swallow back in order to return to the Earth? If you ask such question, their expression would be no less puzzled. But these questions are based on two main theories of physics of the twentieth century. Do we need such theories if they have nothing to do with practice? The question is rhetorical.

We dealt with the one-way paradox and the external force paradox. TEH proposes a radical solution: there is no external force making bodies attract but there is an interaction that can have many manifestations, the basic ones being the movement of bodies toward each other, away from each other, or movement at stable trajectories reflecting a balanced state of the system of interacting elements. When we call the interaction a "two-way street" we need to remember that this street is full of participants and it is an n-body, not a two-body system. Two-way means just a basic vector of interaction: attraction or repulsion. If we consider these bodies as energy oscillators on all scales, from their visible macro parameters to hidden micro processes, we understand that it is all about frequency-phase interactions that can lead to certain phase portraits of the system.

As we have noted in the previous volume, these portraits can be simple or complex. Even simple phase portraits of the Solar system (planetary orbits) are not ideal circles as they were thought to be for centuries. The reason is that the planets are not abstract mathematical pendulums but physical quasi-harmonic oscillators. Their phase portrait would naturally deviate from an ideal limit cycle (circle). Kepler's laws and Newton's corrections that describe elliptical orbits are also not precise. The reason is that the motion of the planets is not a trajectory, which can be described by a single line of some form, but an attractor of a phase portrait, around which the process of the real movement of the phase point oscillates. It can be uneven due to perturbations that affect the phases of the oscillations. These deviations can be of such scale that we cannot observe them in the interval of temporal measurement available to us. For example, the time of Pluto's displacements is expressed in millions of years, and they can be extrapolated and modeled only in computer simulations (Sussman, Wisdom, 1988).

But it does not mean that we should dive into the world of tensors that describe phantom curves. We can provide a detailed description of trajectories using complex geometry. That may give us a better post hoc evaluation but cannot serve as a basis for predictive calculations. For prediction, we need a math that reflects the physical mechanism. Fortunately, a mathematical apparatus for describing systems consisting of many coupled oscillators exists and has solutions for n-body problems. Unfortunately, it is limited to areas outside of mainstream interest in fundamental interactions. It is even worse: any "baroque variations" to mainstream dogmas are prohibited. But there is no need to make variations to these dogmas. Let them live in their own illusory world that is not connected to the real world in any way. What we have to do is to use math that has been developed for describing oscillators and their coupling (more on this in further chapters). In this sense, TEH does not have to invent new math for fundamental interactions but certainly needs help from the existing math for wave interactions because this model is based on the hypothesis that they are the same thing.

From this perspective, we can look at the third paradox which we call the weakness paradox. Gravitation is called the weakest of all fundamental interactions. We can look into any textbook or encyclopedia and see the same statement repeated everywhere. They say that it is so fast, big, and omnipresent that it is hard to "catch it by the tail," so it is weak. That is a paradox, isn't it? The interaction that is moving not only falling leaves and elevators but also planets, stars, and galaxies is called the weakest one. The actual reasons for such a strange situation are the following. First, it is impossible at the current level of our measurement facilities to evaluate gravity in a small laboratory experiment. Second, historically all attempts to measure gravitational interaction proceeded from the wrong assumptions about its nature. But things change and experimenters are looking into gravitation as waves. There is even a project called LIGO — a laser-interferometric gravitational-wave observatory.

The name speaks for itself: the technology is about measuring waves. But what is the problem with gravitational waves? They are very long or low frequency. To hear them we need big ears. LIGO observatory is a colossal complex that is just a huge version of the Michelson interferometer. On the "shoulders" of this L-shaped system, resonators are mounted on special suspensions sensitive to changes in gravity. The incoming wave lengthens one shoulder and shortens the other. The resulting change can be converted into a signal. We can even make a sound signal out of it and hear the "voice of gravity," as the experimenters did.

The experiment is not banal from a technical point of view. Still, from the point of view of the wave model, it merely says that the energy perturbation during the interaction of bodies propagates in the form of a wave. It is the ABC of wave processes. No one is surprised that a clap of palms generates a sound wave. The collision of massive space objects can be called a universal clap. The question is only in the parameters of the source, the propagation medium, and the receiver. The sources are everywhere. The receiver has been built and registered the "clap." Now here comes the major question that concerns the omnipresence paradox. How can an interaction propagate as waves as according to Einstein there is no medium filling all the space?

An outsider may wonder why don't physicists just forget the dogma and follow what reality confirms in an experiment: there are waves so there is a medium for their propagation. Not that fast. Dogmas exist for the believers who are usually deaf to reality if it contradicts the "singular prescriptions" of their prophet. Moreover, they can say that if things are the way they are then the prophet said that this is the way they should be even if this contradicts his actual messages and revelations. This is what happened to the results of the LIGO experiment. It was proclaimed another "test" of Einstein's theory. As with other "tests," this is a substitution of concepts and a trick. Sometimes, it is a plain deceit.

Here is an example: "General relativity predicts that energy can be transported out of a system through gravitational radiation ... The first direct evidence for gravitational radiation was measured on 14 September 2015 by the LIGO detectors ... This observation confirms the theoretical predictions of Einstein and others that such waves exist. It also opens the way for practical observation and understanding of the nature of gravity" (Wikipedia "Gravitation").

First, GTR does not predict that energy can be transported by gravitational radiation as gravitation within GTR is not radiation but the illusion caused by curves in space-time fabric. Second, GTR is based on STR that comes from the first step of abolishing any medium for radiation and postulating the existence of empty space with no physical characteristics. Radiation in the void is physical nonsense. Thus, both STR and GTR do not predict gravitational radiation. On the contrary, they say that it is impossible. But the believers are not interested in physical sense and actual predictions of the prophet written in black on white. For them, the prophet is always right. Here is an example: "Gravitational-wave astronomy is an emerging branch of observational astronomy which aims to use gravitational waves (minute distortions of space-time predicted by Albert Einstein's theory of relativity) to collect observational data" (Wikipedia, "Gravitational-wave astronomy").

Long waves of interaction spreading through vast distances in an all-encompassing medium and measured by a huge interferometer become minute distortions of a phantom in the void. Astronomy which tries to get back physical meaning and measure waves turns out to be a branch of the religion speaking about curves of space-time fabric. Could it be that this phantom is real and astronomers finally discovered it? No way. The article's author lists the observed objects: binary systems of stars rotating close to each other and being sources of pulsations; supernovae; rotating neutron stars and other massive objects. The discoveries are direct evidence of the interaction of large-scale oscillators as a source of ultra-low frequencies propagating in a medium. It's not about geometrical phantoms in the void.

But the prophet is always right: "Gravitational waves have a solid theoretical basis, founded upon the theory of relativity. They were first predicted by Einstein in 1916; although a specific consequence of general relativity, they are a common feature of all theories of gravity that obey special relativity" (Ibid). Everything obeys the dogmas of the prophet.

Here we need to stress that calling Einstein the prophet is not a personal label with good or bad connotations. It aims to show that his attempt to create a theoretical model for physics has become a metaphysical concept with untestable and thus irrefutable postulates that one can only believe or not believe. For some, it is a matter of pride and praise. For example, the propagandist of Einstein's views David Bodanis wrote: "In almost all religions, there's a powerful difference between a priest and a prophet. A priest merely stands below an open hole in the sky, and lets the truth that's normally kept hidden up there come pouring down. (Press secretaries and nuclear technicians are examples.) A prophet, however, is someone who manages to journey up through that opening. They are individuals who can venture to that Other Side, before returning back to ordinary life, here with us on Earth. As a result, we'll try to glimpse, in the expression on their face, or in the potent equations they've plucked and brought back down, what things are like up there, in that higher realm, which so many of us believe in, but know we'll never get to visit directly" (Bodanis, 2000). But for a true scientist, the need to believe in some "higher realm" revelations of a prophet and become a priest of his religious dogma is a point of aversion.

We need to remember that Einstein was just a man who had his insights and happy thoughts some of which turned into doubts and not-so-happy thoughts. Few people know that during his life he changed his views on his own models many times and by the end of his life canceled his own "first step" and the following ones. He effectively invalidated all postulates of STR and GTR.

In his lecture at Leiden University in 1920, which he called "Ether and the Theory of Relativity" Einstein said: "Action at a distance is only apparently immediate action at a distance, but in truth is conveyed by a medium permeating space ... Thus the endeavour toward a unified view of the nature of forces leads to the hypothesis of an ether ... More careful reflection teaches us, however, that the special theory of relativity does not compel us to deny ether ... To deny the ether is ultimately to assign that empty space has no physical qualities whatever. The fundamental facts of mechanics do not harmonise with this view ... The recognition of the fact that "empty space" in its physical relation is neither homogeneous nor isotropic, ... has, I think, finally disposed of the view that space is physically empty. But therewith the conception of the ether has again acquired an intelligible content, although this content differs widely from that of the ether of the mechanical undulatory theory of light" (Einstein, 1920a).

Einstein regretted that, having refuted the mechanical models of the ether, he hastened to declare anathema to the very idea of the presence of a conducting medium. In the same year, in an unpublished article "The basic ideas and methods of the theory of relativity, presented in their development," he wrote: "My opinion in 1905 was that one should no longer talk about the ether in physics. But this judgment was too radical. Rather it is still permissible to assume a space-filling medium ...The theory of space (geometry) and time no longer represent intrinsic physics propounded independently of mechanics and gravitation" (Einstein, 1920b).

Thus, he canceled the basic postulates of both Special relativity and General relativity theories. But it was too late: the myth broke away from its creator and began to live its own life since its followers had their own economic and political goals.

But canceling the void was not enough. It was necessary to create a model of the medium and interactions in it. Although many opponents of the ether still mindlessly repeat Einstein's postulates, with which he began his scientific career, the author himself spent the rest of his life trying to combine interactions and their description into a general model, which he called the "unified field theory." This is how he set the task in 1920: "Of course it would be a great advance if we could succeed in comprehending the gravitational field and the electromagnetic field together as one unified conformation. Then for the first time the epoch of theoretical physics founded by Faraday and Maxwell would reach a satisfactory conclusion. The contrast between ether and matter would fade away" (Ibid).

Did he manage to create such a theory? No. Einstein refuted his views but did not come up with any alternative. This is a complete failure and tragedy for any theoretical scientists. He was just a man who made a mistake, not a prophet who received "otherworldly truths from the higher realms." But he eventually managed to get rid of his own phantoms.

It is worth mentioning here Einstein's last book "Ideas and Opinions," which we can call a confession (Einstein, 1954). The prophet had long doubted the "truths" he had spoken and admitted it. Still, the TR church acquired its own independent life. Few people paid attention to the prophet's attempts to eliminate cognitive dissonance and return physical sense to theoretical physics.

What happened to the postulates about the presence of emptiness and the entity of space-time fabric? He canceled them in one fell swoop. "Space time is not necessarily something to which one can ascribe to a separate existence, independently of the actual objects of physical reality. Physical objects are not in space, but these objects are spatially extended. In this way, the concept empty space loses its meaning" (Ibid).

What about the central dogma, the absolute speed of light? It turns out that, unnoticed by everyone, it has already been canceled. "(Special relativity is founded) on the basis of the law of the constancy of the velocity of light. But the general theory of relativity cannot retain this law. On the contrary, we arrived at the result that according to this latter theory the velocity of light must always depend on the co-ordinates" (Ibid).

An excellent way to get rid of phantoms: first, one phantom cancels the other, and then it just disappears like a dream or an obsession. But what about the postulate about the corpuscular nature of light, quanta-particles, photons-phantoms? What about this black cat, or rather the light cat in the dark room of the twentieth century's theoretical physics? It is not so easy because it was a reason for the Nobel Prize and a fetish for many researchers who stubbornly searched for and even counted the number of these phantoms. It is so popular that any description of light in textbooks and encyclopedias includes statements that light consists of particles-photons. Anyone who begins to doubt this hypothesis is bound to be ridiculed, as most people think it is not a hypothesis but a fact of light's nature. But it was not so evident for Einstein at the end of his life.

"All these fifty years of conscious brooding have brought me no nearer to the answer to the question, 'What are light quanta?' Nowadays every Tom, Dick and Harry thinks he knows it, but he is mistaken" (Ibid).

Einstein sees mistakes of the others repeating his own mistakes. And it's a rare case when he is right: today, the corpuscular paradigm is considered the only version of reality, but it is a mistake.

"The special and general theories of relativity, which, though based entirely on ideas connected with the field-theory, have so far been unable to avoid the independent introduction of material points, … the continuous field thus appeared side by side with the material point as the representative of physical reality. This dualism remains even today disturbing as it must be to every orderly mind" (Ibid).

For any orderly mind that has retained critical thinking, common sense and reality testing function, it is obvious that elementary particles are phantoms of the error of objectification, creating entities "material points" from the discreteness of measuring continuous processes. But Einstein never solved this riddle. Perhaps his only achievement was that he came to understand the fallacy of the paradigm, to which he devoted his whole life.

"What appears certain to me, however, is that, in the foundations of any consistent field theory the particle concept must not appear in addition to the field concept … The great stumbling block for the field theory lies in the conception of the atomic structure of matter and energy" (Ibid).

TEH is based on the same idea that the corpuscular model of the world is the stumbling block for building a unified fundamental interactions theory. But there is also a major stumbling block: the idea of an empty space that leads to the invention of phantoms that can provide for interaction at a distance in this emptiness. If we read carefully his final confession, we see that Einstein nullified his own models: the corpuscular model of light based on the idea of quanta as actual material points (particles); special relativity based on emptiness and an absolute speed of photons flying in it; general theory of relativity based on space-time fabric independent of physical reality.

He turned to physical and common sense which he had laughed at in his young days. His writing acquired an intelligible content at the end of his life. He admitted what was obvious to any orderly mind: wave-particle duality is a myth; continuous oscillatory and wave phenomena which we measure discretely (quantize) is the reality; one unified conformation of electromagnetic, gravitational and other interactions can only be achieved by a model based on an all-encompassing medium; speed of any interaction as a propagating wave (including light) is a variable parameter that depends upon the source, the medium, the observational frame of reference and the coordinate system.

Unfortunately, that was not enough. He did not offer anything to substitute his old models. This produced a conceptual void. That is why the parishioners of his church did not pay attention to his confession. They even called it the rumblings of a senile man. The irony is that when Einstein finally began to make sense and abandoned his delusional models, the parishioners of the STR-GTR church decided that he was talking nonsense.

To acknowledge the mistakes and dead-ends is necessary but not enough. We need to search for a way out of the maze. We have to solve the puzzle of fundamental interactions. It is, of course, vital to get rid of the paradox when an interaction is everywhere but does not have a medium to propagate through. This is the first step to be taken on the road to a satisfactory theory. The next step is to solve the puzzle of the mechanism. For this, we should get over the wave-particle duality paradox. TEH offers a clear solution: there are no elementary particles as material points but only oscillations and waves of an energy environment. This leads us to the idea about the interaction mechanism that seems so trivial in hindsight that one may wonder what was all the fuss about. Oscillations of energy form all levels of matter according to the laws of frequency-phase coupling (synchronization).

This not only helps us to unite all observed interactions in one model based on the mechanism which is universal for all oscillations but to include the model of medium in it. Moreover, we understand that the word "medium" (intermediary) is not precise and even misleading. There is no medium that is separate from interacting objects. There is an all-encompassing energy environment, which is both objects and an intermediary between them. It creates these objects using frequency-phase coupling of vibrations. It also ensures their interaction through its vibrations and waves.

It is useless to look for "entrainment of the ether" by moving matter or to think that it is motionless and does not take any part in its movement. Einstein was right when he wrote in 1910 that the experiments of Fizeau and Michelson-Morley led to a crisis in the mechanical model of the ether. But his conclusion that they confirmed the absence of a medium became the most unhappy thought of his life. If the medium and matter are a single energy environment, then it is impossible to find an "ethereal wind" as a manifestation of a certain displacement between the ether and the Earth as different objects, which is what experimenters tried to do in the 19th century. They didn't find the black cat in the dark room, not because it wasn't there, but because they had the wrong concept of what a cat is. The experiments only indicated that one of the hypotheses about the nature of this environment was not confirmed. In this sense, such a negative result was very important for science. However, the erroneous conclusion about the presence of emptiness was the fall of physics as a science and the rise of metaphysics.

Here we need to look into the history of ether models. In essence, they were mostly mechanical models that were based on the same "good" old corpuscular view of the world: if everything consists of parts (particles, material points) then the medium should have the same structure. Waves turn from the propagation of energy oscillations into the propagation of particle interactions. They become purely mechanical phenomena.

According to Descartes' views, ether consists of primary particles interacting with each other in circular motions, grinding off their corners and turning into increasingly circular shapes (vortices). They touch each other, and the gaps are filled with debris. The density of such a medium allows the interaction to propagate at a tremendous speed. Newton wrote about an elastic and active all-pervading medium. By elasticity, he meant not density but activity. Thus, a very rarefied but active medium can provide a higher velocity than a compressed and inert one. But he also considered it a cluster of particles "exceedingly smaller than those of air, or even than those of light. The exceeding smallness of its particles may contribute to the greatness of the force by which those particles may recede from one another, and thereby make that medium exceedingly more rare and elastick than air" (Newton, 1704).

We face the paradox of any mechanical and corpuscular model of ether. Whatever density we ascribe to it there are still gaps. It is the same problem as with the corpuscular model of matter: when we get down to the scale of the particles, we understand that there is nothing in between. We get into the same stumbling block of interaction at a distance. What is in between? The medium needs the medium. To stop this infinite regress mechanical models inevitably come to the notion of the void. Newton wrote that there could be no interaction in emptiness, but then he wrote that "from the regular and very lasting motions of the planets and comets … it is manifest, that the heavens are void of all sensible resistance, and by consequence of all sensible matter" (Ibid). This legacy of ambivalence about non-emptiness in emptiness (or emptiness in non-emptiness) has passed into the 20th century. Einstein tried to cut this "Gordian Knot" by getting rid of the notion of ether once and for all. But as we have seen this only led him into a dead-end and a need to return to the question of a medium.

Edmund Whittaker, in his book "A History of the Theories of Aether and Electricity," remarked: "It seems absurd to keep the name "vacuum" for a category with so many physical properties, but the historical term "ether" is best suited for this purpose" (Whittaker, 1910). Almost a century later, Nobel laureate in physics Robert Laughlin put it this way: "The word 'ether' has extremely negative connotations in theoretical physics because of its past association with opposition to relativity. This is unfortunate because, stripped of these connotations, it rather nicely captures the way most physicists actually think about the vacuum ... But we do not call it this because it is taboo" (Laughlin, 2005).

The problem is not even that it is a political issue in a "scientific" community that turned into a religious corporation defending its dogmas and proclaiming ideas that run counter to them a taboo. The major problem is the intrinsically corpuscular and, consequentially, mechanical view. The critics are right: "The mechanical qualities of the aether had become more and more magical: it had to be a fluid in order to fill space, but one that was millions of times more rigid than steel in order to support the high frequencies of light waves. It also had to be massless and without viscosity, otherwise, it would visibly affect the orbits of planets. Additionally, it appeared it had to be completely transparent, non-dispersive, incompressible, and continuous at a very small scale" (Wikipedia, "Luminiferous ether").

Any attempt to mechanically model ether leads to insoluble contradictions. But putting a taboo on thinking about the solution leads to even greater contradictions and loss of physical meaning in the leading theories full of phantoms in the void. These riddles are solved only by modeling this medium not as a set of discrete "pieces" but as a continuous environment of energy oscillations (we can call it field or any other name). The properties of such an energy medium are physically plausible. We do not need to give it contradictory characteristics and call for magic to reconcile them. It does not have to be rigid to provide for high frequencies and enormous speeds of propagation. Moreover, as an oscillatory process, it cannot be rigid. It does not have to be some matter without viscosity to provide for the movement of objects in it. Movement in such a medium is just an interaction of different structures of fundamental energy vibrations. We can call it a superconductor where resistance tends to zero due to a highly coherent structure with parameters that allow the establishment of effective synchronization. Thus, it becomes "transparent" for streams of fundamental interaction waves. They pass through it with amazing speed and almost no resistance. The same goes for bodies, as they are also wave structures within this all-encompassing energy environment.

The problem with contradicting characteristics came from the idea that ether is an entity separate from matter. The solution is simple: energy environment and matter are the same thing. The medium is not between something. It is everything and exists in everything. It is continuous on any scale and has no gaps of "empty space." Discrete forms of matter are just some of its manifestations. It is not transcendental to matter but immanent to it. The void exists only in our mind that can produce a model of reality with any phantom. Canceling the void is the essential requirement for our model to become physically sound and adequate to reality. This is the first step to be taken for solving the puzzle of any interaction and all of them together.

CHAPTER 6

THE NUCLEAR MUDDLE

Knowing how to use something is not the same as knowing how it works.

Marvin Minsky

Humankind has been dealing with the problem of controlled nuclear fusion for over 50 years. Some countries spend vast amounts of money on research in this direction. There is also an international project for constructing the thermonuclear experimental reactor (ITER). The deadlines are postponed, and the estimated budget for expenditures is growing rapidly. It is believed that the costs and efforts should be justified since a successful solution to the problem will fundamentally change a lot: the transition from fossil fuels to a source of thermonuclear energy can mean solving the energy issue, which entails a possible solution to global economic and political contradictions.

Thermonuclear fusion produces about 10 million times more energy per unit mass than combustion. Isotopes of hydrogen (deuterium, tritium) can be the fuel for synthesis. Their source in the oceans is abundant. The fusion reaction is fundamentally different in its safety from the decay reaction used in modern nuclear energy: it is not an explosive process, and there are hundreds of times less or almost no radioactive waste.

But to start the synthesis reaction requires an exceptionally high temperature (hundreds of millions of Kelvin and above), i.e., energy is needed to get energy. This raises the question of the ratio of investments to the result. A magical transformation is required, as in paradoxical tales: you have to know how to free a genie from a lamp so that he becomes your slave. To do this, we need an idea about the nature of this genie. But the genie is imprisoned in a lamp, and all that we know about him is that he is mighty, and there is undeniable evidence of the manifestation of his power. One of them is our Sun, in which a constant fusion reaction takes place.

The paradox also lies in the fact that having freed the genie, we must keep him in the lamp and call him on our own accord. We need controlled synthesis. We learned to create explosive a long time ago: the first thermonuclear bomb was tested in 1952. The power of such an explosion, unlike an atomic one, is limited only by the amount of material and can be any. This evil genie has long flown out of the lamp. We look for a kind and obedient one. But to tame it, we have to understand it.

For the explosion, understanding is not essential. We know what conditions are necessary for the process: ultrahigh pressures and densities, which are achieved by the initial directed explosion inward. In such a technology, everything comes from experience: take a certain amount of such and such a substance, bring it to a supercritical state, and then create pressure and density at the right time. Everything is based on the ad hoc and post hoc approach. But using does not mean understanding, and understanding means an entirely different level of use.

The lack of a real theoretical base creates a gap between the phenomena used and their explanation, and this carries the risk of a schism of the model of reality and reality itself. When we talk about this energy level, schism means maladaptation and risk for humankind's vitality. We need not only to survive but to live as best as possible. Desires determine opportunities. If we want to translate desires into reality, our model of reality must be adequate, i.e., provide an explanatory and predictive basis.

What is the genie model currently describing? In general, this is the same old electrodynamics interspersed with elementary particle physics, i.e., the Standard Model. The process is defined as the fusion of lighter nuclei into heavy ones as opposed to a decay reaction in which light nuclei are obtained from heavy nuclei. The paradox is that both are accompanied by the release of energy. Both are called a nuclear reaction, i.e., the interaction of one nucleus with another nucleus or particle, accompanied by a change in the nucleus' structure and composition. Energy output is described as the emission of various elementary particles. The potential energy contained in the nucleus passes into the kinetic energy of these particles' flight, which is called the energy of the nuclear reaction.

Radioactivity was discovered in 1896 by Antoine Becquerel when he studied the effect of luminescence. His idea had nothing to do with particles. He wanted to check whether the emission of electromagnetic waves occurs during luminescence. He was particularly interested in the recently discovered frequency spectrum: X-ray radiation (the range between ultraviolet and gamma radiation). His experiment did not speak of particles. He took the uranium salt, shining in yellow-green light, lit it with sunlight, wrapped it in black paper, put it on a photographic plate in a dark cabinet, and then developed a plate and saw an image of a piece of salt on it. Thus, the hypothesis was confirmed that in this spectrum there were waves of a certain frequency range that could penetrate black paper. It would seem that the result should be recorded as the goal of the "Wave" team.

But several years passed, and the experiments of the father of the nuclear era, Ernest Rutherford, in which approximately the same thing happened, but in different energy ranges, were treated exclusively as particle motion processes.

Rutherford and his colleagues covered radium with a lead container, surrounded it with gold foil, and set up a screen with a layer of zinc sulfide crystals that glow when interacting with certain substances. It is a complete analogy of Becquerel's experiment: there is a radiator, a protective layer and a receiver. The only difference is in the materials and energy level.

What happened? Through a microscope, the researchers observed flashes (luminescence) on the screen. Such flashes were discovered by William Crookes back in 1903, and they were called scintillations. Even though the word carries connotations of oscillatory and wave dynamics, it is believed that each scintillation is the result of the action of one particle. This is what "confirms" the detection of elementary particles in all scintillation detectors. The essence of scintillators is that these substances are capable of emitting light when interacting with external radiation. They are both receivers of waves and sources. The interpretation depends on the point of view. If the discrete manifestation of such a wave interaction is called a particle, then goals are awarded in favor of the "Particle" team.

Rutherford's experiments were such virtual goals. The observed flashes were explained as the effects of individual particles. The picture was as follows: the particles passed through a thin layer of metal, the majority did not deviate, but some deviated at significant angles up to 180°. In the "billiard" picture of the world, such a phenomenon indicated a collision of balls and repulsion from different angles. Why do they repel? The usual explanation was the presence of different charges. But the observed angles indicated that the popular Thomson atom model was not valid. According to it, a positive charge was distributed throughout the atom, and with this configuration of the ball, it cannot push another ball back. Rutherford was surprised but found a way out: the famous nuclear and planetary model of the atom arose. In this model, the atom is filled with void, the center has the maximum positive charge and almost the entire mass, and electrons with a negative charge fly in their orbits.

This was an atomic model based on Coulomb's law. The nucleus and electrons appeared to be those ideal point charges in the void, attracted to each other due to the difference in type of charge. The analogy with the solar system was obvious. But the same question arose: why the electrons do not fall on the nucleus? Bohr proposed a planetary model in which electrons rotate in allowed "native" orbits, and when moving from orbit to orbit, they emit or absorb photons. The explanation of experimental data was simple: the particles flew through the void, collided with a massive and positive nucleus, and scattered at different angles. The interpretation was logical but required the existence of a non-existent, i.e., emptiness.

The interpretation of the expansion of the interaction traces as a natural and physically explainable phenomenon of wave dispersion in a medium was not even considered. This accurately repeated the story with Newton's explanation of the dispersion of light in a prism by the presence of multi-colored corpuscles. Everything would be fine, but such an explanation came up against insurmountable obstacles in the form of other wave effects of light propagation.

But, despite the contradictions with reality and the objections of the "Wave" team members, for example, Huygens, the corpuscular paradigm has become the leading one and has existed for centuries.

As a result, the ordinary and quite peaceful interaction of waves and their dispersion in the medium in Rutherford's experiments was interpreted as a collision of particles in the void and acquired a metaphor of particle war, bombing, and destruction (a coincidence with the ongoing world war is not accidental). The effects of attraction and repulsion were explained by the tautological circle: charges interact in this way because they are of such a type, and a type means that they interact in this way. No one considered the process in terms of the interaction of oscillations and waves in different frequency-phase relationships.

Bjerknes experiments have already been forgotten. The balls analogy was simpler than the hydrodynamic analogy. Contradictions were swept under the rug. Why don't the electrons fall? They cannot, because it is not allowed. Why is the atom structure stable? Because there are native orbits, it is convenient for electrons there: home sweet home. Why, in some cases, the system becomes unstable and even collapses, transforms into another? These are particle life laws: they can fly in orbits, jump, collide, emit, be absorbed and be born. They can do anything, even create phantoms squared: virtual particles produce other virtual particles.

What does it take to destabilize a system? Bomb it with particles. A very intuitive and visual picture, especially when you consider that in the same years (the first decade of the twentieth century) military aviation developed, and the first bombers were created. Is it a coincidence that the first use of nuclear reaction was the bombing of Japanese cities by American aircraft? The metaphor of physicists about particle bombardment of each other turned into the reality of people bombing each other. We often don't even suspect how our metaphors affect our lives. If we look at all wars as the fight for resources and realize that controlled nuclear fusion has almost unlimited energy resources, the change of metaphors that we use to describe the nuclear world becomes not just a mere difference in terminology but a vital necessity. Not only will it help us understand the actual physical mechanism of harmony but it can give us a chance to harmonize our own existence.

So, according to the interpretation of the "Particle" team, the manifestations of interaction observed in Rutherford's experiments mean the bombardment of some particles by others and the destruction of their home, which leads to a forced flight of local residents, accompanied by "spitting out" of photons and radiation. How do we understand that we hit the target? By traces. How do we know that these are traces of particles? We photograph these traces in a Wilson chamber or similar device, and we will call different traces as different particles. Goals in our favor will begin to pour as from a "horn of plenty."

Hydrodynamic analogy came in handy, but only in the interpretation of the "Particle" team. Thus, the "liquid drop model of the nucleus" created in 1936 by Bohr, which underlies modern models of nuclear reactions, suggested that the nucleus can be represented as a spherical drop of unique matter, which has the property of incompressibility, saturation of nuclear forces, evaporation, surface tension, crushing into small drops (nuclear decay) and the merging of small drops into a large one (synthesis). You can describe this with the nuclear binding energy formula called the "semi-empirical Weizsacker formula." It includes such variables as the number of nucleons (mass number) and protons (charge number).

Why is it semi-empirical? On the one hand, it allows us to describe the observed phenomena and is suitable for analyzing some nucleus properties. On the other hand, it is so simplified that when trying to attribute it to these phenomena, there is a need for numerous corrections, i.e., the introduction of arbitrary auxiliary variables. But this is already a familiar trick for mathematical physics.

It turns out that the nucleus consists of identical nucleons interacting with each other, and the binding energy is proportional to the number of nucleons. But if it is a drop, then surface tension means that the outer part contains nucleons that are less connected with others (evaporation). A correction must be introduced to prevent such spontaneous evaporation, otherwise, nucleus will vanish like a phantom. There is also a problem with the charge. According to the model it is determined by charged particles (protons). According to Coulomb's law, they repel each other, which means that the binding energy decreases. To prevent such a spontaneous spreading of the nucleus' charge, it is necessary to introduce corrections into the formula. Otherwise, it is not clear how the nucleus exists at all.

The introduction of a neutral particle (neutron) is also used to explain stability. What does it give? The following explanation appears: the formation of a neutron-proton pair is energetically more profitable than the proton-proton and neutron-neutron pairs. If a balance of balls is violated, there will be an increase in Coulomb repulsion and a decrease in the binding energy (proton-neutron asymmetry). It is necessary to introduce an amendment into the formula to prevent misalliances among the balls.

Empirical data says that the system can be stable, but it can go through bifurcation points and change its phase portrait: either decay or go to another level. To overcome problems in balls' personal life (excess of one gender, shortage of the other), an arbitrary auxiliary variable "nuclei parity" is introduced. It tidies everything up: finds couples for everyone and can even allow harems. And after all such amendments, the formula miraculously (actually, by sleight of hand) turns from semi-empirical to empirical.

The stability of the system and its organization's principle was explained by the "happy and profitable alliances" of balls. The bombing of target nuclei causes the destabilization. The release and emission of energy is explained by the retaliation shots of particles from the nucleus. The decay of a heavy nucleus leads to fragments in the form of other nuclei and by-products of such a "divorce" due to hostilities. By-products are the sought-after energy for which nuclear technology is being created. To cause a "divorce," you have to bombard and undermine the family foundations.

Nuclear fusion in such a model becomes the fusion of particles in "love ecstasy," the creation of native homes with convenient orbits. Such creation requires a powerful source of external energy. Still, the game is worth the candle because the particles, creating new alliances, are so happy that they generate an even more powerful surge of energy in the form of virtual particles that become real radiation. This idea underlies the problem of controlled thermonuclear fusion.

All this is logical and intuitive. With one small correction that makes all these interactions phantom: they occur in the void. The initial premise (such an atom model begins with it) makes all further logical chains meaningless. Formulas can describe anything we want: both reality and phantoms. The supply of arbitrary variables is potentially infinite, so any phantom will be described. Nuclear reactions are written as the sum of the initial particles on the left and the sum of the obtained particles on the right: the balance of virtual balls.

There is one more nuance: real interaction phenomena violate the laws of interaction of particle charges in the classical interpretation by type of charge. On the one hand, all these wonderful and exciting interactions of point charges in the void are described by Coulomb's law. It also "explains" why it is necessary to spend enormous energy on synthesis: nuclei have the same charge and do not want to come together. The Coulomb barrier hinders them. On the other hand, the fact of the formation of structures is evident. This means that there is such a strong convergence of particles that strong fundamental interaction forces come into play that will overcome the power of electrostatic repulsion.

What is this strong interaction that is acting in the opposite direction and violating Coulomb's law? This is again the classic auxiliary variable: the force that had to be invented to explain what binds the nucleon particles in the atom. Here is the mainstream explanation: "Before the 1970s, physicists were uncertain as to how the atomic nucleus was bound together. It was known that the nucleus was composed of protons and neutrons and that protons possessed positive electric charge while neutrons were electrically neutral. By the understanding of physics at that time, positive charges would repel one another, and the positively charged protons should cause the nucleus to fly apart. However, this was never observed. New physics was needed to explain this phenomenon. A stronger attractive force was postulated to explain how the atomic nucleus was bound despite the protons' mutual electromagnetic repulsion. This hypothesized force was called the *strong force*, which was believed to be a fundamental force that acted on the protons and neutrons that make up the nucleus. It was later discovered that protons and neutrons were not fundamental particles, but were made up of constituent particles called quarks. The strong attraction between nucleons was the side-effect of a more fundamental force that bound the quarks together into protons and neutrons. The theory of quantum chromodynamics explains that quarks carry what is called a color charge, although it has no relation to visible color. Quarks with unlike color charge attract one another as a result of the strong interaction, and the particle that mediated this was called the gluon" (Wikipedia, "Strong interaction").

They call it "new physics," but it is the same shell game when balls appear from under a shell when and where a trickster needs them. They can have any color he likes, though it's not the color but just another rule of the game invented by the White Queen as soon as the game runs not in her favor. This is how virtual balls and the forces of their interaction, one more fundamental than the other, multiply.

The process of nucleon interaction was explained by the exchange of particles (pi-mesons or pions). Of course, the experimental "confirmation" of particles' presence as traces in the detectors came in a while. They began to explain attraction or repulsion by one nucleon emitting a pion and another absorbing it, just as the electromagnetic interaction is described by the exchange of virtual photons.

Since the repulsion increases inversely to the distance, one must try very hard to bring the nuclei closer and activate a strong nuclear interaction, working at distance scales of the order of the nucleus itself. It can overcome repulsion only in very tight contact. Probably, otherwise, the nucleons cannot reach each other so that one emits a pion and the other swallows it. You have to put them in a closed chamber. For example, in a toroidal chamber of a Tokamak or another container of a thermonuclear reactor. Next, you must give them great kinetic energy, which will break the Coulomb barrier. Then the miracle of the exchange of pions and the conception of a "sun child" of thermonuclear energy will occur.

The barrier is so high that the substance involved in such a reaction will be a very energetically charged plasma. Another detail is interesting: a neutron is most interesting for attempts to use thermonuclear energy. It has a low Coulomb barrier and is "light on his feet": a strong bond fastens the newly formed nucleus, and neutrons fly out and carry kinetic energy with them. We will only have to catch them in our networks and use them (preferably for peaceful purposes).

But in the description of all nuclear reactions, a reservation inevitably arises: specific energy and spin ratios are required. What are these ratios? In the billiard paradigm, energy ratios become the number of balls and their charges. In the 1950s, particles began to multiply at high rates. They had very short lifetimes but were strongly interacting. They had different spins and charges. Some regularity was visible in their distribution, but where it came from was not known. By analogy with pion-nucleon scattering, a model of the strong interactions of these hadrons was constructed, in which each type of interaction, each type of decay corresponded to its own interaction constant.

In addition, some of the observed dependencies could not be explained, and they were simply postulated in the form of "rules of the game," which hadrons obey (Zweig rule, conservation of isospin and G-parity, etc.). The description seemed to work but certainly was unsatisfactory from the point of view of theory: too much had to be postulated, a large number of free parameters were introduced entirely arbitrarily and without any structure. It means that the model is a failure. The author of the Wikipedia article puts it in a gentle way: "Some physicists consider it to be *ad hoc* and inelegant, requiring 19 numerical constants whose values are unrelated and arbitrary" (Wikipedia, "Standard Model").

The model has neither explanatory nor predictive power but is smart in the art of producing equations in which the number of virtual particles and their mysterious quantum properties are adjusted to the result of energy measurements. The hierarchy of spirits is so complicated that no diagrams help. SM cannot explain the regularities that are manifested in interactions without creating virtual and quasiparticles with special quantum properties. It means only one thing: there is no understanding of the mechanism "under the bonnet." We need not such "new physics," but only physics, which would remember that its task is to explain and describe material phenomena and not create mathematical tricks with the phantoms' participation.

Regularities and patterns of energy interactions were seen in all experiments. There were attempts to explain, but along the usual beaten path of creating virtual particles. Hadrons were invented as a class of particles subject to the strong interaction. To describe the observed patterns, hadrons were assigned quantum numbers (strangeness, charm, beauty, etc.). Then hadrons had to be divided into quarks. Depending on the composition, hadrons became known as baryons (three quarks of three colors or colorless combination), mesons (one quark, one antiquark). Nucleons that make up the nucleus (proton, neutron) and hyperons (unstable particles) were assigned to baryons. Pions, kaons and other virtual entities were assigned to mesons.

The quark concept was at first a purely mathematical construction for describing the degrees of freedom (parameters) of hadrons. According to the law of reproduction of entities in the billiard model, it was "discovered" that these are particles that carry hadron momentum, charge, and spin. Hadron became a composite particle. The problem was that no one had observed any quarks but continued to divide particles into particles to explain the empirical facts and the laws of energy relations in the continuous physical process with different levels of amplitude-frequency characteristics and phase portraits. Mathematical abstraction, according to the established tradition, has become an entity. Moreover, this entity was called a fundamental particle, but with the provision that it does not exist in a free state, therefore it is impossible to observe it. Spirits always behave this way: they exist, but they are not to be found; we can only believe in them.

In the 1970s, they created the theory of interaction of quarks in accordance with their quantum number called "color" (quantum chromodynamics, QCD). Although the quarks themselves were postulated as some kind of pointless structureless objects, they had to have specific properties to explain the interactions. For different manifestations of the interactions, varieties of quarks (flavors) and their properties (colors) were needed. For some reason, these point and fundamental particles had a fractional charge, and they began to be divided into types known as flavors: up, down, charm, strange, top and bottom. It is Wonderland full of quark spirits. The names are not accidental: the metaphors accurately reflect the model, which is very strange for everybody but still, charming for the believers.

There are no physical explanations, but phantoms arise instead. These spirits are omnipresent and omnipotent. Quarks are involved in all interactions (strong, weak, electromagnetic, gravitational). What happens to them? According to QCD, strong interactions (gluon exchange) can change a quark's color but do not change its flavor. Weak interactions, on the contrary, do not change color but can change the flavor. The unusual properties of strong interaction lead to the fact that a single quark cannot move away from other quarks at any significant distance, which means that quarks cannot be observed in free form (a phenomenon called confinement). Only hadrons as colorless combinations of quarks can fly away.

The quark coupling decreases proportionally to the distance, and the quarks flying nearby were considered noninteracting (asymptotic freedom). But at the same time, they were not free, but captive: they could not move farther from each other than the size of the "prison cell" (confinement). But they can outwit the jailers, merge and form a colorless quark, and in a colorless state, they can run as far away as if they are invisible. The result is that in the real world, there are only colorless combinations of spirits that are identified with hadrons.

Runaway quarks can interact, not through the exchange of gluons, but of other hadrons, for example, pions. Then it is a strong interaction and holds nucleons in nuclei. Antiquarks were required to explain the opposite vector of interaction. A classical scheme of the annihilation of a particle and antiparticle was offered to explain the neutral state. And as usual, a special field appeared: calibration field that describes the interaction of quarks. Quanta of this field were called gluons. Each gluon has its own independent field and interacts with the color of the quark. Gluons also have color and interact with each other. The theory says so, but no one has ever observed any quarks or gluons in any experiments.

Since these are classical auxiliary variables, they work to hide contradictions under the rug. They become unique entities that can violate the rules established by previous versions of the model, where the set of entities was smaller. They are "free parameters" just for that purpose. They have quantum states that do not fit into the fundamental particles' rules. They have fractional charges: 1/3, 2/3 of the electron charge. We recall that the assumption of the presence of an elementary charge, of which all charges consist, was put forward in the eighteenth century by Franklin, and experimental "confirmation" was found at the beginning of the twentieth century when the idea of quantizing any variable became very popular. The elementary charge was called the "minimum portion" (quantum) of charge, the fundamental constant. But then suddenly, it turned out that the elementary charge is not at all elementary but divisible.

We can measure anything in any way. We can quantize (sample) in any way and create scale units for this. But when we begin to give our measurement the character of a "fundamental constant," then inevitably, sooner or later, with a further increase in our measurements' resolution, we will encounter contradictions. This happened with the elementary charge, and it was necessary to come up with virtual particles with remarkable properties.

In addition to these, another virtual entity was added: a magnetic monopole, as an elementary particle with a point magnetic charge; a source of a static magnetic field, just as an electric charge is a source of electric field. Do not forget that the fields multiply with the particles. They are united in alliances (for example, an electromagnetic field), but they are still spoken of as separate entities interacting with each other. If there should be an elementary charge with a certain sign, then there must be a magnetic monopole, i.e., single pole entity. The problem is that all the observed phenomena are binary (dipoles), and no one has ever observed a monopole.

The discrete billiard ball model requires a mono-structure, and the continuum of energy reality does not want to be such, and stubbornly demonstrates the fluctuations between yin/yang. But the apologists' hopes do not die because the belief in the presence of at least one monopole confirms another belief in the quantization of all electric charges in the Universe. But the observed phenomena of energy interactions each time lead to contradictions with the model, and entities with the right to violate the rules arise.

Where did the concept of fractional "elementary charge" come from, which then turned into a property of newly invented particles? Experiments showed that under certain conditions (ultralow temperatures, strong external energetic effects), an abrupt change in the conductivity of a solid material occurs. The phenomenon of a change in conductivity and the appearance of a transverse potential difference when a DC conductor was placed in a magnetic field was discovered at the end of the nineteenth century (Hall effect). The integer step-like change in new experiments was called the "quantum Hall effect." The dependence turned out to be a multiple of the accepted measurement of the elementary charge. It was so stable and evident that even a new standard unit of resistance had to be introduced.

What was the reaction of the "Particle" team? Standard and within the Standard Model framework: it means that there is one more (or not one?) type of particles with part of the elementary charge. The real laws of the interaction of oscillations are ignored, and virtual particles are regularly born in the minds of theorists. There was even the concept of anion (from the word 'anything'), as new particle that became useful for describing the fragmentation of the process. Literally anything goes to save the model.

Subsequent experiments showed that the ratios could be 1/3, 2/3 and 2/5. Anion type statistics arose and violated the principles of fermion and boson statistics. They even started talking about anion interference (an oxymoron with a complete loss of physical meaning). The "Particle" team stubbornly ignored the obvious: the presence of integer-valued interactions of different levels of frequencies of energy oscillations and wave patterns.

The fractionality of the charges of quarks and gluons is only the beginning. They also violate Pauli's ban on sharing a single quantum state. This is another fundamental principle, which turns out to be completely non-fundamental. Recall that this prohibition was a necessary theoretical construction in Quantum Field Theory and SM to justify the presence of electron shells in an atom and explain the whole variety of chemical elements. The planetary model of the atom required fixing the native orbits based on at least some principle. Electrons have become fermions, which are simply forbidden to be in the same quantum state, and therefore they fill different levels (orbitals) sequentially. Do not forget that all the electrons are the same by the definition of the model, so it is in no way possible to find out exactly which quantum state a particular electron is in. For the atom model to somehow correspond to the observed phenomena, the principle just has to work. As for the electrons, they will figure out who is who by themselves.

It is usually believed that there is no analog to such a principle in classical physics, and there is a familiar reference to the special strangeness of quantum physics, which does not have to be understood, but simply taken for granted. Wave models that can explain the distribution of energy levels and their interaction are not considered by the "Particle" team as a matter of principle. When reality violates the principles prescribed in the theory, the team's main principle stays intact: everything should happen to particles and be explained only by particles. Thus, unscrupulous particles preserve the main principle of the team. New quantum characteristics are invented that allow quarks to be in any state in any combination. This successfully "resolves" some of the contradictions. Since the new abstraction helps the model remain afloat, it inevitably becomes objectified and turns into another object that no one has ever observed. What was a mathematical description of certain parameters, becomes an object. Then it is possible to ascribe a spin and an orbital angular momentum to it. You can draw the rotation patterns of these balls relative to each other and thus explain the spins and momentums of composite particles (hadrons).

Here a miracle happens: no need to change the model (except for the violation of some principles by particles without principles); no need to rebuild anything (other than adding new virtual entities); experiments "confirm" the model (more precisely, virtual entities explain observations post-factum); we can explain the presence of observed divisions of "fundamental quantities" and the presence of jets during high-energy interactions as decay of composite particles into virtual parts as a result of collisions of balls in the void; the new virtual parts themselves begin to multiply and interact with each other with the help of new virtual particles (gluons); magical transformations begin, births and especially the favorite annihilation of particles and antiparticles by each other, the formation and decay of pairs (synthesis and splitting).

But questions remain. Why precisely three colors? Why exactly three generations of quarks? What are quarks made of? How do quarks add up to hadrons? They are again attributed to the unresolved problems of modern physics. In fact, it is one and the same question about the mechanism of interaction and the formation of the structures of matter.

With gluons, of course, the story is the same. Nobody has ever observed them anywhere, but they explain everything wonderfully, albeit with violation of previous principles. They are simultaneously carriers of interaction, i.e., should be fundamentally neutral, but they are the owners of the "color" and interact with each other without any principles. They are true spirits, like photons: they have neither mass nor charge. There are colorless gluons, which are antiparticles to themselves, have no charge, momentum, isospin, strangeness, charm, color, and can explain the interactions of other particles by their existence that has no physical characteristics. Non-existing entities as the basis of all the interactions. A never-ending story in quantum Wonderland.

What is considered "evidence" of the presence of such particles? First, the convenience of their characteristics for the model. Second, indirect evidence in the form of the appearance of hadron jets. Why are traces of real wave energy processes in the medium left on the calorimeter in the collider considered evidence of the flight of virtual particles without mass and charge in the void? See point one: the presence of particles is a convenient description, even if they are three times virtual.

How are the characteristics of these virtual particles measured? By breaking up the calorimeter into small cells, producing highlighted energy, which is defined as the energy of particles colliding with the calorimeter's material. The discreteness of the measurement of the continuous wave process becomes an object called a "particle." The formulas are adjusted to the measurement result, and this is considered proof of the fidelity of the model. Such is the main unscrupulous principle. It is beneficial for career growth and obtaining titles and awards but has nothing to do with physics.

If there is a logic in all these transformations, exchanges, captures, liberation, swallowing and emitting the visible invisibility of colored and colorless phantoms, then there is only one: this is a desperate attempt to explain the observed energy processes by creating entities. A classic mistake of objectification of the process and going into endless regress, when each created entity requires the creation of a new one to explain the old and oneself, and so on and so forth.

Specific amplitude-frequency characteristics of oscillations, frequency-phase relationships during their interaction, dynamics and changes in phase portraits, combinations of different frequency levels, all polyphonic and polyrhythmic music of matter turns into some kind of chaos of particles with strange relationships wandering in the void. The charm of these relationships is debatable. It is a matter of aesthetic taste. But there is no physical meaning there. What does this state of theory lead to?

Apologetics claims that all the most accurate experiments in which SM's descriptions were tested confirmed its predictions. As we have already seen, predictions actually have zero force since when contradictions arise, the next renormalization or invention of the virtual entity, which responsible for correcting the contradiction, occurs. But the set of tricks does not end there. There is another, less costly, but no less effective: banal ignoring the facts and results of experiments that do not fit into the theory.

For example, in the 1970s, Alan Krisch and colleagues conducted experiments to identify patterns of interaction of proton beams polarized in one direction of the spin (Krisch, 1979). The technologies for reducing energy wave fluxes in accelerators to one polarization have been mastered for a long time. Still, in the billiard paradigm, such a beam is represented as a set of balls with rotation axes pointing in one direction (spin polarization). The experimenters created a single polarization of the flow and the fixed target in order to check the results for different spin directions. When the spins' directions coincided, the scattering occurred at the flux and the target, as in unpolarized beams. Such a picture fits into the model of the collision of balls and their flight in different directions. The scattering of waves in a medium could be mistaken for the expansion of particles in the void. One could get away with ignoring the oscillatory and wave patterns. But then there was an embarrassment: with opposite polarizations, the scattering cross-section dropped sharply, i.e., given the level of accuracy of the polarization technology, it could be said that scattering did not occur. Stunned, Krisch wrote that it looked as if the particles passed through each other. The results were, to put it mildly, unexpected for the billiard model. They did not fit into the new stream of SM named QCD. Under the predictions of the model, they simply should not exist.

Some adherents tried to give explanations within the framework of QCD, but, as the father of the theory of electroweak interactions and Nobel Prize laureate Sheldon Glashow said, the experiment turned out to be "the thorn in the side of QCD" (Krisch, 2005). There was another option to neutralize the result: to declare that it is a mistake due to equipment deficiency. But experiments repeated by Krisch and other researchers in the 80s and 90s on more advanced equipment with a degree of polarization close to one hundred percent showed the same results with greater accuracy (Peterson, 1990).

So, the spin direction played a fundamental role in the interaction (spin effects). In one case, dispersion did not occur; in another it occurred in different directions with different concentrations. This did not coincide with the predictions of the model. What should be a normal reaction? Correction of the old model, if possible, or the creation of a new one that can accommodate the entire class of observed phenomena, rather than "picking cherries." The attempts to adjust failed, but the second option was unacceptable to apologists of dogma. The works of Krisch were no longer mentioned in scientific journals. Following the behests of the founding prophets, the believers swept the contradictions under the rug. After all, this is a proven way to get Nobel Prizes.

The results were explained as a manifestation of "elastic scattering," in which particles do not change their structure during interactions. They were declared "less fundamental," and SM, they say, is engaged in fundamental interactions (inelastic and changing the composition of particles). The balls collide elastically and then can scatter in different directions as they want or slip through each other. When they collide not elastically, their structure changes, there should be no spin effects, and various magical transformations of particles into each other with the exchange of virtual constituent particles and uniform scattering come into play. But what about the description of the first case of the collision? What about the fact that inelastic collision spin effects also occur? This does not fit into the usual and canonized particle exchange diagrams. This is a taboo. It is dangerous to speak about it even in a whisper. But what if we forget not about real phenomena but about incorrect models and their dogmas? It is blasphemy. For this, you can get on auto-da-fe.

Suppose we pluck up courage together with common sense and take the side of the "Wave" team. We begin to develop hypotheses about the spin as a phase portrait of oscillations with specific amplitude-frequency characteristics, sequence and direction of phases. In that case, the polarization ratios become phase interactions of the oscillations, and all spin effects become physically explainable. No stunning stunts of fundamental particles flying through each other. No thorns in our side. The effect becomes the natural phenomena of waves passing through each other. No miracles, but in a way banal thing that anyone can imagine as it is common to all observable wave interactions. Throw two stones at the same time into the water, and you can watch how the waves calmly do what is impossible for particles, even if they are three times virtual. The phenomena of dispersion, interference, and diffraction are universal for high energies and everyday experience. There is only one condition for these commonplace phenomena to occur: the presence of an environment instead of emptiness.

As an element of the medium, an oscillation can participate in many waves at the same time. The superposition of different waves, as the addition of oscillations at each point in the medium through which these waves pass, depends on the frequency-phase relationships. The coincidence of frequencies and phases can lead to a literal passage of waves through each other without scattering. The difference in phase velocities and frequencies causes dispersion, which can be very different and does not have to be distributed evenly. Exactly these different effects were demonstrated by the experiments of Krisch and other researchers. The same "proton spin polarization" is one phase and frequency, different — antiphase at the same frequency. In the absence of any clear polarization, there is other phase coupling. It is physically causal and explicable but only for waves in the environment. For phantom particles in the void, it is just another twist in the weirdness of the quantum Wonderland.

But if you think that indivisible particles passing through each other is the ultimate weirdness, wait till you hear about the "quantum Cheshire cat." We all remember that Alice was puzzled by many things in Wonderland. But one phenomenon stood out: the cat that could dematerialize leaving only its smile behind. When artists try to illustrate the phenomenon, they inevitably draw lips without a cat. But that is not what the author wrote. Lewis Carroll definitely wrote about the cat leaving a smile, not lips. Otherwise, it would not be a Wonderland. A smile is a property of lips being in the process of smiling. It is not an object that can have a separate existence. The process cannot be separate from an entity where this process is happening. Only if we make an objectification error and think of a smile as some object, can we separate it from the cat. Thus, the picture of lips separated from a cat is an example of how our common sense, remaining within the framework of physical reality, tries to cope with the riddle of the abstract idea of a smile separated from a cat.

What does this all have to do with the quantum Wonderland? It has the same riddle. Usually, it is described like this: a paradoxical phenomenon in quantum mechanics, the essence of which is that a quantum system under certain conditions can behave as if particles and their properties were separated in space. As usual, it is claimed to be the result of an experiment. However, we should be very attentive to the actual experimental setup as the interpretation of a result depends upon how we interpret what we measure in the experiment.

First, let's take a standard description of the experiment. A beam of neutrons passes through a silicon crystal and divides into two parts. The first neutron beam has a spin along the neutron path, while the second beam's spin is oriented in the opposite direction. After going through different paths, both beams combine and produce an interference pattern. Then, the direction of the spins is slightly changed using a magnetic field. The model predicts that a change in the spins should lead to changes in the entire interference pattern. During the experiments, it turned out that the magnetic field applied to the first beam did not produce any effect. If a magnetic field is applied to a second beam, the desired effect appears. Suppose we interpret beams as flying particles with spin as some rotational property. In that case, we inevitably interpret the observed result as a quantum Cheshire cat: particles fly in their direction, and their spin flies in the other direction. If you think that is a weird conclusion, you have not parted with common and physical sense yet. But this is not what quantum physicists are famous for. They are famous for never adhering to any meaning. Here is the interpretation of the results by the authors of the experiment: "The experimental results suggest that the system behaves as if the neutrons go through one beam path, while their magnetic moment travels along the other" (Denkmayr et al., 2014).

But let's remove from the description of the experiment any references to phantom particles and talk in the language of the physics of wave processes. A wave with one polarization (direction of the oscillation vector) passes through a crystal and divides into waves with opposite polarizations. It is the physical wave effect of a change in polarization due to refraction and reflection when passing through an anisotropic medium. We remember that the properties of the waves depend upon the medium of propagation. So, no wonders so far. When these waves interfere with each other the pattern depends upon their properties, including the phase vector direction. So, nothing separates from its properties as far.

When we send other waves to interfere with the original ones, we inevitably influence the resulting pattern. Please, note that we do not use the term "magnetic field" as it creates another phantom. We adhere to the physical meaning and talk of waves as waves, not particles flying in the void and fields acting on them. So, one wave is sent to interfere with the other two waves. It is a basic configuration of a physical experiment: the dependent variable is manipulated using an independent variable to determine how the first one is influenced by the second. In this case, we mix waves with certain properties with other waves to see how those properties change or don't change. The results should not surprise us if our model is correct.

When interacting with a new wave with the same polarization, one of the initial waves did not change the direction of the oscillation vector. It was the natural effect of waves passing through each other (preserving the initial parameters) at certain frequency-phase relationships. When interacting with a wave with a different polarization, the second wave changed its phase portrait. As a result, the interference pattern of two initial waves in one case remained unchanged, and in the other case, it was changing. So, the experiment was actually about the phenomena of the interaction of waves with different frequency-phase coupling variations. The experiment just confirms the usual behavior of waves that can change their properties in the process of interaction in the medium. No miracle of separation of properties from their carriers occurred. Not a single Cheshire cat left behind a smile. Physical reality is dull. To be precise, a realistic story is not a fairy tale but a model of reality that has physical meaning.

Here is an opinion: "The possibility of separating a particle from one of its intrinsic properties ... is rather intriguing and questions a very basic everyday notion, by which the properties of things are always with the things themselves ... The basis of the controversy lies in the attempt to attribute physical reality to a situation that simply cannot be perceived as physical reality ... The only mystery left is the usual quantum mechanics weirdness, in which particles can be detected individually, while their propagation satisfies wave-like properties" (Corrêa, 2015).

The conclusion speaks for itself: when "individual detection" (quantization, a discrete measurement of a continuous wave process) is taken to be the existence of a quantum as a discrete object (particle), then such an error of objectification is the starting point of departure from physical reality and the source of all weirdness of quantum Wonderland. This leads to the situation when practical physicists have two options: either to interpret the results of their experiments within the leading theories' dogmas about phantoms and part with physical meaning or to forget about the theories and take the results as some unexplainable mysteries. It is interesting to note, that in both cases the interpretation will sound like a description of a miracle. The only reason for this stalemate is the absence of a physical model.

Evgeny Velikhov, Chairman of the ITER Council, said: "We still do not understand why we manage to maintain a temperature of 100 million degrees in Tokamaks. It is generally a miracle. It would seem that plasma physics is based on simple laws. There is no need for the theory of relativity, no need for quantum mechanics. Everything can be calculated on the basis of Newton's law and Maxwell's equations. But you cannot do this on the most modern teraflop computers" (Velikhov, 2003).

Indeed, if the leading models do not explain anything, but refer to wonderful fields, virtual particles and fabrics, then a practical result is a miracle that no one understands. Einstein was not at all joking when, in response to a question about the relationship between theory and practice, he said: "Theory is when you know everything, but nothing works. Practice is everything works, but no one knows why. In our lab, theory and practice are combined: nothing works, and no one knows why." We have to remember that he did not have any lab and based his models on thought experiments. Indeed, in his theoretical work, he made every effort to combine theory and practice so that nothing worked, and no one knew why. But life goes on, and practitioners have to dodge and do something while continuing to be at a loss as to the results.

Leading physics theories are useless, although they claim the privileged position of universal truths that do not require adjustment. In practice, something works, but as an incomprehensible miracle. The statement of the academician sounds like this: we do not understand what we are doing. For a person who is far from physics, this will sound as "a bolt from the blue," since there is a myth in the mass consciousness that all these amazing nuclear technologies are based on the understanding by physicists of the processes that they are engaged in. For those who know the real state of affairs, such an acknowledgment is a statement of the fact of daily practice: leading theories and models created in the twentieth century have nothing to do with reality. The Nobel Prizes for theoretical models are simply stamps with the words "approved" on the next description of a mathematical trick. What is the outcome? The fundamental questions about the structure of matter and energy interactions are unresolved.

John Clauser, an experimental physicist, wrote about the founding fathers of QM: "The untidy legacy left by the above authors was particularly acute, since neither side attempted to justify its position with hard experimental data. It was automatically assumed by them that, without any doubt, quantum mechanics is "obviously" the correct theory … Thus, it seems that all of these authors believed in what I will call here the "standard religion" … Correspondingly, the theory needs no further testing, even in areas where its predictions may seem to be surprising and/or paradoxical … Totally mathematical approach leaves an experimental physicist in the very awkward position of having no specifically defined experiment (or set of experiments) on which quantum mechanics is purportedly based! … Perhaps these authors all tacitly accepted and promoted the aforementioned religious belief because they were all theorists and felt correspondingly they should act as clergymen. At the very least, they clearly show negligible respect for experimental physics as the final arbiter in physics … Religious dogmatism then quickly promoted a nearly universal acceptance of quantum theory and its Copenhagen interpretation as gospel, along with a total unwillingness to even mildly question the theory's foundations … Given the omnipresent religious zeal, it would seem to be sheer folly for anyone to propose new experimental tests of quantum theory. What if their results disagree with quantum theory's predictions? We certainly will all perish! … The net impact of this stigma was that any physicist who openly criticized or even seriously questioned these foundations (or predictions) was immediately branded as a "quack." Quacks naturally found it difficult to find decent jobs within the profession … Any student who questioned the theory's foundations, or, God forbid, considered studying the associated problems as a legitimate pursuit in physics was sternly advises that he would ruin his career by doing so. I was given this advice as a student on many occasions by many famous physicists" (Clauser, 2002).

Here is an interview with one of such students who took the advice not to question foundations and just left for specialized practical research where he could avoid dogmas of the clergymen: "When I was young, physics seemed to offer answers to all of the mysteries of the universe. It felt authoritative and unequivocal in its explanations of nature and the origin of the universe. In that sense it was the perfect religion for my teenage self … Those books were always so dogmatic like the Catholic Sunday school I went to as a kid … Theoretical particle physics is definitely a dead subject. Other areas of theoretical physics have made great strides in applications but at the same time there hasn't been any fundamentally new development in our understanding of physics for decades" (Christopher Search, blogs.scientificamerican.com/cross-check/whats-wrong-with-physics).

But practice is impossible without a theoretical concept. In this situation, practitioners have to rely on ancient models of classical mechanics and electromechanics only for the lack of better models. This does not mean that these old models correspond to the current level of penetration into the depths of matter and the need for its explanation. But at least they describe reality and do not go into an endless abstraction of mathematical phantoms. Newton and Maxwell's equations, which modern practical physicists use, work to a certain degree of approximation and averaging. They do not explain anything but describe the manifestations. Moreover, they are useless for describing energy oscillatory processes in nonlinear systems. This explains why even modern computers cannot give the correct calculations when using these formulas. We can increase the teraflop power of a computing device as much as we like, but if it was given the wrong model for calculation, then the result will be wrong. Trite, but true.

We can look at what is happening in Tokamak from the wave model perspective and, probably, it will not seem such a "miracle" anymore. One of the main problems is the stabilization of the process. For the reaction in Tokamak to be controllable, it is necessary to keep plasma in a limited volume. No walls can keep the thermonuclear reaction with its temperature within controlled limits. Therefore, in the 1950s, technology using electromagnetic interactions was proposed. When describing the process, the usual terms are used: the plasma's electric current in the toroidal chamber creates a magnetic field, which holds the plasma in the system. At present, plasma stability time is only seconds. This, of course, is not enough for such an installation to be called a controlled thermonuclear fusion reactor.

The second problem: there is no synthesis yet, since the temperature, although "wonderful," is insufficient. To achieve a result in the form of a thermonuclear reaction, huge currents are required, and the resource is not infinite. What are they trying to do? Introduce additional injections of different feeding options (deuterium, tritium, and others) into the chamber. But heat due to the flow of current in the system itself is not enough for a thermonuclear reaction. There is one exciting option for additional heating: the effects of external microwave radiation at resonant frequencies with the cyclotron frequency. Practical physicists go in the paradigm of wave and oscillatory processes. Perhaps, one way or another, they will be able to "tune the pipes," and a real and understandable miracle of synchronization will occur.

Let's look from the new concept perspective at such seemingly incomparable objects, like a colossal Tokamak, where they try to free a "genie" of enormous energy from a "lamp" of a toroidal chamber with a plasma, and a small container with water, where thermonuclear energy levels arise from interactions with sound. For clarity, let's take an analogy of breaking a glass with voice. All that is needed for such a "miracle" is to find out the glass's natural frequency by clicking it with a fingernail and singing so that the frequency of the voice resonates with it. As a result, the glass may break due to an increase in the intensity of vibrations. In all of these examples, we are talking about energy oscillations and their phase-frequency coupling.

Even a result of minimal stability indicates that there is a frequency-phase coupling in the Tokamak system. But further selection of parameters is necessary and, possibly, as in the sonoluminescence effect, tremendous energy efforts are not required to free the "genie." You have to slowly rub the lamp with the right frequency and in the proper phase. Or, as in the glass analogy, click the glass and listen to its music. Fine-tuning of pipes is needed, and miracles become real. Moreover, they cease to be miracles. They turn into understandable and even ordinary technologies.

There is another "miracle," which, out of habit, was first called force: a weak nuclear force. Here it must be emphasized once again that the use of the word "force" has two categories of meaning. The first category means the measurement parameter of a signal and process. In this case, it is synonymous with many words: power, amplitude, intensity, quantity, mass, potential, influence, voltage, strength, hardness, stability, pressure, prevalence, etc. These are physical quantities that can be measured and evaluated. They are empirically recognizable, and the use of the word "force" has a strictly physical sense. In this study, when describing processes, it is used with such meaning. But there is a second option, which comes from the error of objectification and turns the process and its parameters into an entity called "force."

The tradition of using this word in physics has both sides. The second one is a legacy of the ancient animism and religious worldview, explaining all the phenomena and processes by guiding transcendent entities. An implicit understanding of the unacceptability of such use in scientific discourse is manifested in the fact that gradually it is being used less and replaced by the word "interaction." But centuries-old traditions do not pass without a trace, and often, when difficulties in explaining physical phenomena arise, it emerges in some theories as an auxiliary variable. When describing such concepts, it is necessary to put it in quotation marks to emphasize the similarity to the "holy spirit" and going beyond the scope of science.

In SM, the word "force" was gradually replaced by interaction, but in the corpuscular model's traditions, it is invariably interpreted as the interaction of elementary particles. Is weak interaction different from others? We recall that to explain the binding of nucleon particles in an atom, a special strong interaction was invented, acting at distances of the scale of the nucleus in the opposite direction from electrostatic repulsion. This force got its name, fields, carriers, and the usual mechanisms of emission, absorption, birth and annihilation of particles. But the question arose: why does the spontaneous decay of nuclei occur? After all, some substances are naturally radioactive. Interestingly, the term means "creating waves" (from lat. radius — ray, activus — effective). It was coined at a time when the "Wave" team still had the right to speak.

So, is there a specific force pushing particles away? It is logical: if there is an attractive force, then there must be a repulsive force. The question is in the mechanism. The billiard model says clearly: radioactive decay is a spontaneous change in the composition or internal structure of unstable atomic nuclei by the emission of elementary particles, gamma rays or nuclear fragments. In any case, the formulation is discrete and speaks of particles. Why are nuclei unstable, and these fragments fly apart? The answer is the following: a weak but fundamental force responsible for nuclear fission lives there and acts on the scales of distances even smaller than the nucleus, even smaller than the diameter of the proton part of the nucleus. But what is the mechanism of its action? To explain, the theory of quantum flavordynamics, similar to quantum chromodynamics, appeared. Later it was replaced by the theory of Electroweak Interaction, which was considered more successful and unifying. The authors were awarded the Nobel Prize in 1979 "for their contribution to the unified theory of weak and electromagnetic interactions between elementary particles."

Let's go back to the experiments of Becquerel, who discovered the phenomena of radioactivity. Rutherford did subsequent experiments and decided that traces signified particle collisions, which he called alpha and beta particles. Furthermore, many researchers' experiments have shown that beta particles deviate in a magnetic field and the deviation corresponds to the accepted concept of charge negativity. Becquerel showed that they have the same charge-to-mass ratio as electrons. Then it was discovered that the electrons emitted during decay could have arbitrary energy. Where this energy came from was not clear, and birth from nothing contradicted the law of conservation of energy. Besides, the electrons possessed stable phase characteristics, which were designated as spin 1/2. To resolve the contradictions, Pauli hypothesized that there is another kind of particles emitted during beta decay: neutrinos, as neutral fundamental particles with a half-integer spin and related to leptons, i.e., particles not participating in the strong interaction.

As usual, a special force required its carriers. These neutrinos are so small, neutral and weak that they interact very little with matter and can fly through void and matter with a mean free path (distance between collisions) of hundreds of light-years. They can be detected if they are very dispersed (high energy) and begin interacting with the target. On this basis, Enrico Fermi developed the first theory of beta decay, which, by analogy with the already existing idea of quantization of light with emission and absorption of virtual photons, said that particles emitted from an atom were not initially contained in it but were born in the interaction of others particles. Interestingly, "Nature" magazine did not publish his article because of too much abstractness. Then they were still not quite used to such details of the intimate life of virtual particles. But time passed, and soon this abstractness no longer bothered anyone, but, on the contrary, delighted.

In the 1950s, it turned out that something strange happened during decay: the law of symmetry was violated. It was believed that the result of any experiment should be completely mirrored if the conditions and experimental setup of different experiments are mirrored. First, there was a confirmation that the symmetry is not preserved during beta decay, then it turned out that the same happens during the decay of muons and other particles. Such a phenomenon had to be explained. According to the usual scheme, Feynman and colleagues advanced a theory of four-fermion weak interaction (V-A theory), which indicated that vector and axial currents were involved in the weak interaction. The vector current behaves decently (remains unchanged during spatial inversion), and the axial current changes its sign, which leads to parity violation (P-symmetry). They also differ in charge parity: they violate C-symmetry as invariance with respect to the charge conjugation operation (replacing a particle with an antiparticle). In other words, they behave as they want and change their identity on a whim or the need to maintain order in the world.

The currents themselves consist of various virtual particles and are, respectively, lepton, hadron, quark currents of all possible flavors, from charmed to very strange. For all this to combine, "operators" arise, and all the terms in the current are the sum of the vector and axial operators. Combinations are described by "quark mixing matrices" as a change in flavors during weak decays. Quarks have six flavors, and their combinations give compound particles (for example, neutron and proton) their properties. The uniqueness of the weak interaction was such that it allowed the quarks to change their flavor, and carriers of force (bosons) mediated the change. For example, in beta-minus decay, the lower quark in the neutron becomes upper, which leads to the transition of the neutron into a proton and the emission of an electron and antineutrino. The neutron emits a heavy W-boson with a charge of -1 and turns into a proton with a charge of $+1$, and the boson decays into an electron and an antineutrino, as the final fragments of beta decay. At the same time, magic happens: the mass of the W-boson is 100 times the mass of the neutron emitting it.

At first, they tried to preserve the symmetry and said that combined CP symmetry arises (a combination of mirror reflection and particle replacement by antiparticles). But then it turned out that such CP parity is broken due to weak interaction, and new generations of quarks and leptons were required, which entailed their mandatory "discovery" and another Nobel Prize. The culmination of events was the model of electroweak interactions that introduced special gauge fields and quanta of these fields: vector bosons W^+, W^- and Z^0 as carriers of weak interaction. There is no surprise that later, these particles were "discovered." A machine for producing virtual particles has long worked in the field of other interactions. Why should the weak force not keep up with this race? It is weak but fundamental.

What is the bottom line today? "The Standard Model of particle physics describes the electromagnetic interaction and the weak interaction as two different aspects of a single electroweak interaction. This theory was developed around 1968 by Sheldon Glashow, Abdus Salam and Steven Weinberg, and they were awarded the 1979 Nobel Prize in Physics for their work. The Higgs mechanism provides and explanation of the presence of three massive gauge bosons (W^+, W^-, Z^0, the three carriers of the weak interaction) and the massless photons (γ, the carrier of the electromagnetic interaction). According to the electroweak theory, at very high energies, the universe has four components of the Higgs field whose interactions are carried by four massless gauge bosons — each similar to the photons — forming a complex scalar Higgs field doublet. However, at low energies, this gauge symmetry is spontaneously broken down to the U (1) symmetry of electromagnetism, since one of the Higgs fields acquires a vacuum expectation value. This symmetry-breaking would be expected to produce three massless bosons, but instead they become integrated by the other three fields and acquire mass through the Higgs mechanism. These three boson integrations produce the W^+, W^- and Z^0 bosons of the weak interaction. The fourth gauge boson is the photon of electromagnetism, and remains massless" (Wikipedia, "Weak interaction").

Everything is interacting through the exchange of massless particles. At high energies, it is all balanced somehow, but at low energies, something happens spontaneously, and all breaks down. This breakdown should produce other massless spirits but by the wave a magic wand these spirits materialize and become massive. What is the magical mechanism of this birth of mass out of nothing? In 1964, Peter Higgs and Francois Engler independently created theories that became an important part of the Standard Model. They explained the inert mass by the presence of a specific vacuum field (later called Higgs field) that is responsible for "fundamental interactions and provides the Higgs mechanism of generating mass of elementary particles" (Wikipedia, "Higgs Field"). This field produces elementary particles (later called Higgs bosons) with zero spin and no electric charge which are so unstable that they decay into other particles immediately upon generation. Nevertheless, this fleeting ghost with no properties is responsible for the whole massive world construction. For this, it was even nicknamed by physicist and Nobel laurate Leon Lederman as "The God Particle".

How does this God create the world? We do not expect any surprises as we know about the essence of SM: everything is about virtual balls. Here is how it describes the mechanism by analogy. Scattered on the surface of the table, small foam balls spray from the slightest blow. However, being poured onto the water's surface, they no longer move as easily as the interaction with the liquid gave them inertness. The Higgs field is analogous to water and ripples on the surface of the water are analogous to the Higgs bosons. They usually add that the inaccuracy of this analogy is that water interferes with any movement of the balls, while the vacuum field does not affect particles moving uniformly and rectilinearly but counteracts only their acceleration. Thus, mass arises as an inert reaction to acceleration.

The Higgs mechanism turns causality upside down. The usual physical meaning of the concept of inertia is quite intuitive: it is a quantitative property of a physical system that has a qualitative property of material substantiality (mass). Thus, it is not inertia that creates mass, but inertia is a property of massive bodies and one of the primary manifestations of mass. It is not surprising that SM breaks causality and logic. If virtual ghosts (or gods?) are responsible for the creation of the material world, there is no need to adhere to physical meaning. Anything goes in the world of phantoms. However, even when describing phantoms in the void (vacuum field is the new name for it), physicists cannot avoid analogies with interactions in a real physical environment (water). It is the hydrodynamic analogy but in a perverted way. The model speaks of interactions in a medium but has to use the concept of virtual balls in the void. The dogmas hang as the "sword of Damocles" over the heads of mainstream theorists.

There is also no surprise that they inevitably found the ghost with the declared property of having no properties. It is also not surprising that they found it in the so-called "particle collider" which is in fact a device for measuring the interactions of waves of energy. We have already mentioned that the terms and metaphors used by the "Particle" team members not only substitute the actual wave measurements with phantoms of the corpuscular model of reality but reflect the way they think about this reality. Thus, the interaction and harmony of waves in the first experiments with radiation were interpreted as collisions and bombardments of particles.

The same story happened with the so-called colliders. The initial name for all such devices was synchrotron (from synchronization and electron). The principle was invented by Vladimir Veksler in 1944 and it was about wave phenomena, not particle collisions. Modern facilities like the Large Hadron Collider (LHC) and others are in fact synchrotrons. But the name changed. Sync of waves is not within the mainstream flow of thought. They are all into particle collisions: "The synchrotron is one of the first accelerator concepts to enable the construction of large-scale facilities, since bending, beam focusing and acceleration can be separated into different components. The most powerful modern particle accelerators use versions of the synchrotron design" (Wikipedia "Synchrotron").

The device is about controlling the interacting beams of energy (real waves in the medium) but the model is about acceleration and collision of phantom particles in the void. Should we be surprised that there is no practical outcome of such devices? Instead of leading to breakthroughs in providing controlled energy they only use huge energy recourses produced by other means to look for more phantoms: "The aim of the LHC's detectors is to allow physicists to test the predictions of different theories of particle physics, including measuring the properties of the Higgs boson and searching for the large family of new particles predicted by supersymmetric theories as well as other unsolved questions of physics" (Wikipedia, "Large Hadron Collider").

The large family of particles grows without any restrictions, since there is no way to prevent believers from measuring the properties of a ghost, which, according to the definition given by its inventors, has no properties and does not even exist for a period of time sufficient to measure it. If you think this sounds strange, remember that it is an old trick: by proclaiming an immaterial entity, one can become the prophet of a belief system that cannot be refuted.

If there is a prophecy, there should be a fulfillment: "The Large Hadron Collider at CERN announced results consistent with the Higgs particle on 14 March 2013, making it extremely likely that the field, or one like it, exists, and explaining how the Higgs mechanism takes place in nature" (Wikipedia, "Higgs mechanism"). Of course, they called it a confirmation of the theory predictions and Nobel committee put an "approval stamp" by issuing the prize in physics to Peter Higgs and Francois Engler.

But was it about physics? Not at all. It was the same old story about virtual ghosts. SM postulates that the Higgs boson is born from the "intercourse" of two other virtual particles (gluons) and breaks up into virtual carrier particles (quarks, antiquarks, photons) and container particles (electron-positron pairs, and/or gluon-antigluon with a muon-antimuon neutrino pair). It is all about ghosts giving birth to ghosts. Is it physics or metaphysics? Is it a prediction about physical entities or a prophecy about immaterial entities? For a person who has not parted with common sense and thinks about physics as a science that should be dealing with material entities, the answers to these questions are obvious.

The mechanism of imbalance during weak nuclear interactions is described in the Higgs model as a spontaneous violation of vacuum symmetry and illustrated by the usual ball analogies. The ball is first on top of the hill in equilibrium but unstable state, and then rolls down, where the state is equilibrium, but asymmetric (one direction of movement is selected from all possible):

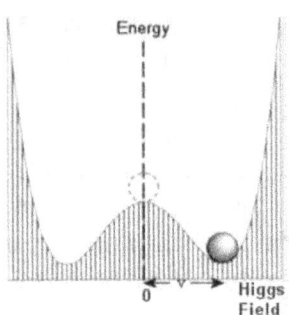

Symmetry is described as a balance of weak and electromagnetic forces when virtual particles (leptons, quarks, bosons) are not manifested. "The breaking of symmetry triggers the Higgs mechanism, causing the bosons it interacts with to have mass" (Ibid). So, when the symmetry of emptiness breaks, the ghosts become manifest and acquire material substantiality (mass). This is the essence of the mechanism of how immaterial entities produce matter. Even ancient religious myths of creation sound more physical. For example, God of the Old Testament produced forms of matter from physical energy: "And the earth was without form, and void; and darkness was upon the face of the deep. And the Spirit of God moved upon the face of the waters. And God said, Let there be light: and there was light" (Genesis 1-4:16).

But within the dark void of the SM religious myth, the light can be seen. As we have noted, all analogies taken to describe the mysterious "vacuum field" speak about a physical energy environment. For example, the postulated quality of the Higgs field is that it has non-zero energy even when there are no forms of matter in it. This makes it a full analogy of the ether concept. The problem remains: the lack of an idea about the physical mechanism that can produce forms of matter from this energy environment makes the model a religious concept speaking about virtual entities. In this sense, it is no different from the Genesis description: it just postulates that some all-mighty spirit produces matter from the void in a miraculous way that cannot be explained.

The description of balls acquiring mass due to contact with the field gives an illusion of an explanation because it does not answer the question of how forms of matter appear. The balls have to materialize before they interact with anything. Higgs mechanism says that it is vice versa: they interact and then materialize. They are the spirits that move upon the face of the water breaking any physical causality.

Moreover, these spirits are unique. They are allowed to materialize and acquire mass because they carry weak interaction at tiny distances. But the photons and gravitons that perform a long-range job have to stay massless. They are not allowed to materialize. But what about gluons, which are postulated as zero mass spirits carrying strong short-range interaction? To get out of this internal contradiction, the rule of confinement was set and these spirits were declared to be even more special. They cannot be observed separately and are always in clumps forming composite hadrons.

But did anyone actually measure all these entities and their mass? As we have mentioned, not a single experiment, even in the devices built specially to test predictions of the Higgs mechanism theory and measuring the properties of the Higgs boson, caught these spirits in the act. There are two reasons for this. First, they cannot be measured as they have no properties to measure. Thus, from the start, it is a trick. Second, the "particle colliders" are actually wave measurement devices with high sampling frequency and quantization. Thus, from the start, it is an objectification error and the search for a black cat in a dark room that is not there. It is no surprise, that when they started to measure interactions (strong or weak, fusion or decay) in energy parameters (electronvolts) and calculated mass according to the standard energy-mass equivalence equation the values contradicted any physical meaning. The bosons turned out to be heavier than some atoms (for example, iron). Thus, a simple fact of significant energy changes became another paradox for the model of flying balls.

How did the Standard Model get out of a contradiction? In the standard way: hiding it under the rug with the help of renormalization which even the inventor called a "dippy process" and a "shell game" (Feynman, 1985). But such a scam has already become so familiar that it is perceived as the norm. Moreover, if the founder got the Nobel Prize for this scam, why shouldn't the followers be awarded? The proof of such a model's renormalizability by Martinus Veltman and Gerardus 't Hooft was awarded the 1999 Nobel Prize in physics. In this form, the theory of weak interaction is included in the modern SM version. The postulated virtual particles are inevitably detected, and the formulas inevitably turn out to be true, since each time they are adjusted to the result of the experiment by renormalization. Meanwhile, the model is farther and farther from reality and practice.

The tradition of calling different amplitude-frequency levels and phase changes of the oscillation states as particles, their spins, charge types, colors, flavors, etc., seems unshakable. The decay of the energy structure and the emission of energy waves of a specific frequency range in an energy medium, accompanied by a change in this structure and its phase portrait, are invariably described as flights, birth, disappearance, emission, absorption, and other life events of virtual particles. Thus, the traditional billiard ball paradigm creates the illusion of explaining the fundamental processes of attraction, repulsion, establishment of balanced combinations and their destruction by the game of spirits in the void.

As a characteristic of such a pathological modeling of reality, the ambivalence is manifested in all aspects. Theorists, on the one hand, understand that a coherent and unified model is needed because the general laws of the process are traced. On the other hand, they remain in a vicious circle of "phantom carousel" of theories declared standard and indisputable. Getting out of this circle means a huge career risk. Moreover, a career in academic institutions is generally impossible without putting on a mask of an adherent of mainstream theoretical physics' church dogma. Cognitive dissonance affects the state of the followers themselves and the state of the entire field. Meanwhile, practitioners are forced to remain in a state of ambivalence. On the one hand, they must speak the generally accepted language of existing dogmas, and on the other, ignore them in work.

But one way or another, physicists are trying to overcome the ambivalence and come up with various workarounds to overcome the taboos and describe an all-encompassing energy environment. For example, the Higgs mechanism's mathematics has roots in the theory of condensed matter developed by Lev Landau and Vitaly Ginzburg. The properties of the Higgs field are analogous to the state of superconductivity described by this theory. But the problem is not that overcoming the heavy legacy of Einstein's abolition of the environment and the taboo on ether proceeds with great difficulty. Let it be "vacuum field." As we have emphasized more than once, the word does not matter. The meaning put into it is crucial. The meaning will come when we understand the physical mechanism due to which the observed manifestations of energy interactions occur.

CHAPTER 7

THE WAY OUT OF THE MUDDLE

When you know that the Great Emptiness is full of Energy, you understand that there is no such thing as Nothing.

Zhang Zai

We need a cardinal paradigm shift. It sounds complicated and grandiose, but it will happen when the necessary and sufficient conditions are created. One of the necessary conditions, maybe even the most essential, is an elementary step: to cancel the void and resurrect the environment. In spite of all the taboos of the mainstream dogmas, many theoretical physicists try to create models that one way or another include an all-encompassing energy environment that they call with different names.

The Theory of Energy Harmony (TEH) does not avoid the word "ether" However, it is not the best term as it has connotations of ethereal as something immaterial. Moreover, this word has many other meanings from the world of technology and organic chemistry. The word "field" carries the echoes of objectification error, as it produces an image of an entity that is transcendent to the occurring phenomena. The term "medium" limits the concept to an intermediary between something. The word "environment" has connotations of something being outside of something, while we are talking about the energy that is both a means for interaction and the source of the interacting forms of matter themselves. For lack of a better term, TEH uses "medium" and "environment" interchangeably to refer to the all-encompassing, continuous, and active system of self-sustaining energy fluctuations that constitute the Universe.

In some cases, these processes and the medium itself are directly observable. But in most of the cases, the medium is not available for direct observation. To make a model that includes such an unobservable entity scientific we should not define it as something unobservable in principle. Otherwise, we are back to metaphysics which speaks of immaterial spirits. It is not a problem if we do not see something. After all, our eyes have developed in evolution for the purpose of transducing signals of a particular frequency range that we call visible light. Moreover, all animals have different abilities for electromagnetic signals processing, and thus their ranges of frequency spectrum vary. For example, the red color does not exist for a dog as its brain does not process the long wavelength part of the spectrum. But it does not mean that there are no such waves. Some frequency ranges are not processed by the human brain but are available for other species. We compensate for our deficiencies and widen the available range by developing artificial signal processing technologies and transducing unavailable ranges into those that are within our senses' capabilities. Thus, signals that we had no notion about "come into existence."

The idea is simple: if some ranges of the environment's signals are not processed yet, it does not mean that it is an empty space without any physical quantities. We can flip it the other way: if there is an all-encompassing energy environment with a potentially infinite range of frequencies, we can widen our range of signal processing to make more of it available for our observation. This assumption does not contradict physical meaning and causality. Moreover, it can and should be a base for the rise of physical science from its fall. Canceling the idea of the void is not a negation but an affirmation. It is the way out of the muddle that theoretical physics has been in for a century.

If we proceed from the assumption that all waves in all frequency ranges are the propagation of vibrations in a medium (in fact, it is the definition of a wave), we can follow this path to explain interactions that happen in a visible medium and in an invisible one by analogy. Let's get back to basics once again and look at the history of studies of wave phenomena.

We can start with Chladni figures. The name of Ernst Chladni is not familiar to the public and even many physicists. The Wikipedia article is concise and contains a minimal description of his work on acoustic phenomena and meteorites. However, we can call him the father of both acoustics and meteoritics. Chladni called himself a wandering artist because to spread his ideas, he traveled around Europe with lectures, demonstrations of experiments and the performance of music with the instruments of his invention. But he had something else to show the public besides unusual instruments. He scattered sand on a flat sheet of metal and vibrated it with a violin bow:

The bow is a source of oscillations, and the metal sheet and sand are conductive oscillatory media. What should be expected of such a system of interacting oscillators? Wave structures and their phase portraits. So, various patterns of coherent structures began to form on the surface of the sheet. Their shape depended on the source's oscillation frequency, on the parameters of the medium (material, shape, elasticity, natural frequencies), and the place of excitation. By changing the parameters, Chladni could show "miracles" creating different forms:

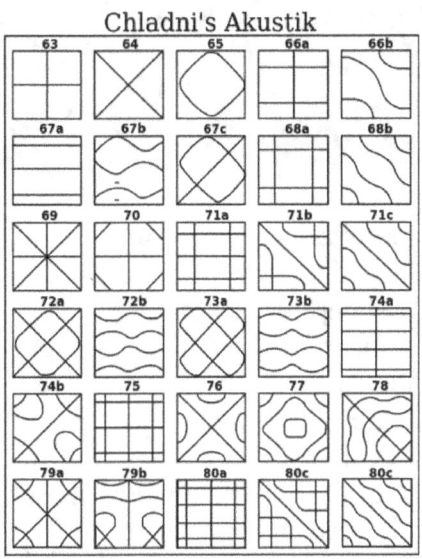

Chladni published his results in a book called "Discoveries about the Theory of Sound" (Chladni, 1787). However, he did not explain the phenomenon, and the theoretical model was postponed for a long time. In the nineteenth century, when the phenomena of wave interference became more or less clear Chladni figures got their first explanation. Waves superpose, and, depending on the frequency and the distance between the sources, amplification or damping of oscillations occurs. The alternation of amplitude peaks and dips forms a line structure. Energy is distributed over space, and the maxima compensate for the minima. In the case of Chladni's experiments, everything seemed simple to the point of banality: sand is dumped from places where the amplitude of oscillations is greater and collected in locations where it is less. It is no accident that, subsequently, such areas of standing waves were called antinodes and nodes. However, it was an illusion of an explanation as the mechanism that produced the specific figures and their dependence upon source, medium, and receiver parameters was not explained. After all, the question is precisely why the wave energy is distributed in such a way that, with certain frequency combinations, specific and stable structures form.

Some researchers continued to look into the phenomenon more deeply. Interestingly, they were most often not physicists, but artists or from other professions. For example, in the 1950-60s, the doctor and artist Hans Jenny studied the effects of sound waves on a wide variety of materials: from powder and liquids to spores of mushrooms and wet plaster. His experiments gave interference patterns in 3D. He invented the tonoscope: a container filled with different media.

For example, a liquid containing suspended particles and randomly turbid, when exposed to harmonic combinations of diatonic intervals, demonstrated the emergence of stable forms. Synchronization created order from chaos. Look at the beauty of the music of matter (Jenny's figures in a drop of water):

As long as there is a synchronized state of harmony, the structure is maintained. As soon as it is broken, the structure either becomes unstable or breaks up. Transitions occurred by jumps after decay periods. It is a clear manifestation of the change in the phase portrait and the transition from one synchronization region to another: at the boundaries, desynchronization occurs; with parameters entering a new Arnold tongue area, a new phase portrait of the synchronized system appears. Defining parameters are amplitude, frequency and phase relations.

Jenny was so impressed by the apparent presence of patterns in all this diversity that he considered this a discovery and called the field of studying the interaction of sound and substances "cymatics" from the Greek κῦμα "wave" (Jenny, 1967). This term did not gain popularity among physicists. Theoretical physics was especially keen on the race for the ghosts of virtual particles in the void, and funding and bonuses to the "Particle" team poured from the horn of plenty. What is the point of listening to eccentrics talking about the harmony of waves in the world when money and titles are there, where there is a continuous war with bombardment and particle collisions in the void? Moreover, these eccentrics are not graduate physicists, which means profane. But the profane people have one advantage: they do not have blinkers on their eyes. Their eyes are free from dogma. They cannot always explain what has been observed, but they don't hide the facts under the rug to satisfy the parishioners of some church called the "leading theory of physics." They are initially outside the mainstream, and they do not need to stay afloat in it.

Chladni showed that with certain combinations of source and medium parameters, wave structures with nodal lines are formed. But in the experiments of Jenny, there was one difference from the experiments of Chladni with sand: for liquids and finely dispersed substances (for example, powder), the formation of structures was the opposite, i.e., they accumulated in antinode zones (the largest amplitude) and escaped from the nodes. It turned out that they were attracted to energetically charged areas, in contrast to sand, which seemed to just fall off from regions of intense vibration. This contradicted the usual interpretation of the figures as simply "energetically advantageous" lines (the substance accumulates where it shakes less). The mystery remained.

There was another inconsistency. Jenny's experiments showed that even with the system parameters' stability (medium properties, source frequency), the forms could have dynamics. Vortex-like structures with continuous motion arose in the liquid. It was the self-organization of a dynamic complex process in action. The figures drew strange attractors with sensitivity to the initial conditions, but with their own dynamics. Moreover, when enthusiasts began to continue experiments in different environments, the resulting volumetric figures started to resemble the ubiquitous fractal forms of inanimate and living matter. This could be called an accident if not for the stubbornness of reality, which "knocked on our door." It is not at all surprising that some of these basic forms corresponded to what atomic orbitals look like if the electron is considered as a wave packet, and the wave function is calculated in accordance with its energy parameters:

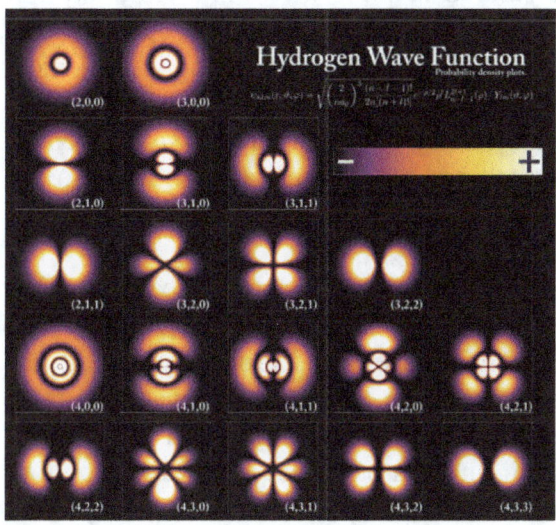

SM interprets these forms as the probability of finding a particle in a particular region of the void. It sounds weird, but this is the essence of the Copenhagen interpretation of the wave function. In Chladni and Jenny's experiments, the presence of the medium is obvious. Its parameters clearly affect the dynamics of the processes. There is no reason to call the physical waves in the medium the mysterious probability of particles flying in the void. If we start talking about wave processes when the medium is not apparent, a stumbling block arises.

Let's go back to the times when classical physicists were also profane and did not have dogma blinders on their eyes. For example, Michael Faraday did not have a formal physical education. He was self-taught, and his university was a bookstore where he worked as a delivery man. He studied the old theoretical models, but they did not create insurmountable dogmatic barriers, since he was free from pressure from authorities and career considerations. Despite his well-deserved fame, he remained an independent researcher until the end of his life. He repeatedly refused to be promoted to chivalry and the post of president of the Royal Society.

He was very impressed with Chladni's experiments for a reason: they had obvious analogies with the subject of his research, electromagnetism. The discovery of electromagnetic induction by Faraday, of course, was a great event. The ability to convert electrical energy into mechanical and vice versa became one of the foundations for the subsequent industrial revolution. But Faraday was interested in the main question: how stable structures are formed as a result of the dynamics of wave processes in the medium. This obvious question has not yet sunk in the dogmas about virtual particles in the void.

The Chladni figures provided a guide for his scientific inquiry. Faraday was not perplexed by the invisibility of the environment. The concept of ether was not considered taboo during his time, as the age of theoretical physics religions had not yet dawned. He examined the environment that was available for direct observation and drew analogies. For instance, he investigated the properties of granular medium under the influence of vibration and observed various patterns forming. He also studied standing waves in a vibrating fluid and observed stable regular structures, which were later called Faraday waves. He discovered that vibrations of plates placed in water caused longitudinal and transverse waves. This was an analogy of electromagnetic phenomena with multidirectional energy propagation. Faraday found that the frequency of perpendicular waves was equal to half the oscillation frequency of the plate (octave ratio 1:2). In general, the form of wave structures depends upon frequency-phase coupling and is stable as long as the ratios stay in harmony.

Today, almost two hundred years later, researchers continue to observe these phenomena, giving them different names. For example, the structures in vibrating granular media are now known as oscillons. Here is the description: "In physics, an oscillon is a soliton-like phenomenon that occurs in granular and other dissipative media. Oscillons in granular media result from vertically vibrating a plate with a layer of uniform particles placed freely on top. When the sinusoidal vibrations are of the correct amplitude and frequency and the layer of sufficient thickness, a localized wave, referred to as an oscillon, can be formed by locally disturbing the particles. This metastable state will remain for a long time (many hundreds of thousands of oscillations) in the absence of further perturbation … Oscillons of opposite phase will attract over short distances and form 'bonded' pairs. Oscillons of like phase repel … The experimental procedure is similar to that used to form Chladni figures of sand on a vibrating plate. Researchers realized that these figures say more about the vibrational modes of the plate than the response of the sand and created an experimental set-up that minimized outside effects, using a shallow layer of brass balls in a vacuum and a rigid plate. When they vibrated the plate at critical amplitude, they found that the balls formed a localized vibrating structure when perturbed which lasted indefinitely. Oscillons have also been experimentally observed in thin parametrically vibrated layers of viscous fluid and colloidal suspensions. Oscillons have been associated with Faraday waves because they require similar resonance conditions. Nonlinear electrostatic oscillations on a plasma boundary can also appear in the form of oscillons" (Wikipedia, "Oscillon").

So, there is an experimental fact that, with specific amplitude-frequency characteristics of the oscillatory medium and its elements, stable wave structures form. They can persist for any length of time provided that the oscillation parameters are unchanged and interact with each other in certain frequency and phase ratios. Such a frequency-phase coupling can manifest itself as convergence (attraction) and divergence (repulsion) of structures. Such manifestations are clearly analogous to the fundamental interactions. But the author of the Wikipedia article does not make any associations with the central mysteries of physics. After reading the article, it is difficult to appreciate the importance of the phenomenon. If a person is not interested in the way patterns in sand form, he will quickly forget about oscillons or Chladni figures.

Even the authors of the experiments did not make any conclusions about the obvious analogies with other types of interactions. They just remarked that the effect could be "important in industries ranging from food to mineral processing, yet a basic understanding of the physical mechanisms underlying the collective dynamics of grains is lacking" (Melo, Umbanhowar, Swinney, 1995). But what was the experiment really about? About the specific dynamics of granules or about the dynamics of any oscillatory medium? About both, since the patterns of oscillations and their interactions are the same.

By manipulating various parameters, the researchers found that the oscillatory dynamics turned out to be decisive. At particular amplitude-frequency characteristics, molecular- and crystalline-like structures in the form of lines, chains, lattices, squares, hexagons, and more complex patterns were established:

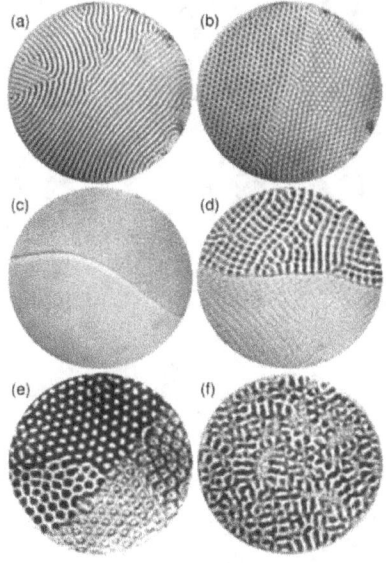

These structures oscillated on subharmonics of the driving frequency, demonstrating the frequency response spectrum of the system. When changing vibration parameters, transitions between different phase portraits occurred through bifurcations. Sometimes several coexisting phase portraits appeared in the system (Fig. E). With a different frequency response, the system came into a chaotic state of numerous wave structures with short spatial and temporal correlations (Fig. F), and then clear patterns appeared again.

The authors noted: "The ordered structures we observe in granular media have similarities with those in vertically oscillated liquids (Faraday experiment), but an important difference is that period-doubling, which is unobservable in fluids, leads to domains with different phase in the granular layer. With this exception, it appears that the factors determining pattern selection in the two media are similar even though ... the effective forcing in the granular layer is different from the smooth parametric forcing of the fluid" (Ibid).

Of course, patterns and their formation depend not only on the amplitude-frequency characteristic of the source but also on the medium's properties. A granular medium is different from liquid or solid bodies and often demonstrates a unique combination of properties inherent in both. It has strong friction increasing energy dissipation, which means that an external driving oscillator has to be strong. Like other media, it has an internal mode of oscillations, as a set of harmonic oscillations with its natural (resonant) frequency and stable phase relations. Due to this property, it can establish frequency-phase coupling in various synchronization modes.

In addition to the above patterns, for a specific range of the driving frequency, stable localized structures appeared, which were radially symmetric forms with oscillations on this frequency subharmonic. There was a stable phase portrait with fluctuations from peak to crater and vice versa:

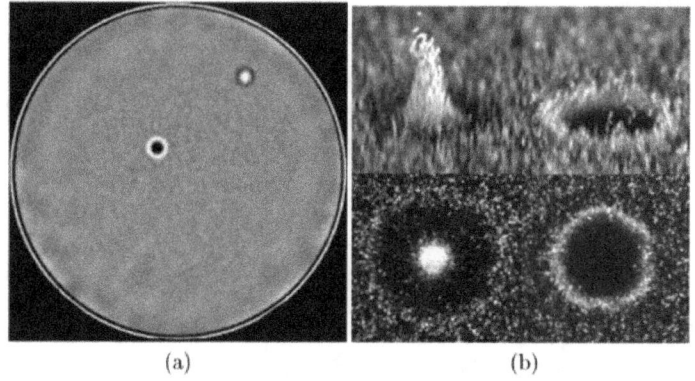

(a)   (b)

Shattuck, Bizon, Umbanhowar, Swift, Swinney, 1998

This is what the authors initially called oscillons. Their oscillation frequency was two times less than the plate's frequency (an octave ratio of 1/2). Thus, a localized structure with the frequency different from the medium (plate and neighboring space) frequency appears within a stable synchronization region. But even more impressive was the fact that the oscillons can move and interact in certain phase relationships. Oscillons in antiphase do not absorb each other during a collision, but retain their structure, attract and form a stable pair. Such pairs are capable of creating chains and other complex systems of molecular and crystalline type. The opposite direction of interaction also works: there is a repulsion of in-phase oscillations. The push-pull balance creates a stable phase portrait of the entire system.

If we continued to think in terms of the charge type, we could say that Coulomb's law works here. But in the presence of an evident environment, it becomes clear that we are talking about the frequency-phase relationships of oscillatory processes in this environment, and not about the mystical attraction and repulsion of charges of a certain "gender." The charge type concept is convenient for describing the manifestations of the process but does nothing for its explanation and understanding of the mechanism. There are no "point charges" with some mysterious property, which we call a positive or negative sign of the charge. There are stable localized structures with certain energy characteristics, phase portraits of oscillations and phase relations.

This approach introduces physical meaning into the very concept of electric charge. It ceases to be a certain quantity determining the ability of bodies to be a source of electromagnetic fields and to take part in electromagnetic interaction. The confusion, whether the fields determine the bodies' interactions or the bodies are the source of the fields, goes away. An electric charge becomes one of the levels of energy oscillations with specific amplitude-frequency characteristics. Different interactions exhibiting similar regularities are then explained by the difference in energy levels with fundamentally the same physics of the process.

The units become truly arbitrary scales that can be converted into each other without any stretch. For example, one coulomb, as energy passing through a cross-section of a conductor with a current of 1 ampere per second, gives the same coupling strength of two charge carriers at a distance of 1 m as the coupling strength of a body of 1 million tons with the Earth. Now such conversion of coulombs into newtons is done based on the similarity of formulas without explaining why the phenomenological descriptions of different interactions turn out to be identical. At the same time, mainstream theories, which supposedly explain the phenomena themselves, speak of completely different mechanisms of electromagnetic and gravitational interaction.

There is no law according to which one charge type is attracted to the opposite. The unexpected manifestation of the same types' attraction does not require another special law, force and field. In this sense, energy fluctuations have the freedom of "gender relations," but they obey the interaction laws. There is a specific combination of frequencies and phases, which can lead to attraction and repulsion. There are no separate laws for bubbles in sonoluminescence, spheres in the Bjerknes experiment, planetary systems, galaxies, electromagnetic, nuclear, gravitational interactions, oscillons, vortices, whirlpools, and other manifestations of oscillatory and wave dynamics, synchronization and desynchronization. There are features of system elements' parameters, their combinations, and phase portraits of the systems themselves. The basic laws of interaction in such systems are the same, even though the environment at first glance is very different.

The discreteness of the medium consisting of granules is as arbitrary as any continuous energy medium's discreteness. We can divide the medium into elements, but until we consider it a single dynamic oscillatory system, we will not understand the laws of the processes. The main mistake of all attempts to create a mechanical model of the medium of fundamental interactions was not even that it was considered as a set of parts. The basic error of the ether's mechanical models was an attempt to create a model of the medium as an entity separate from the structures forming and interacting in it, to separate energy from matter. If we continue to think in the categories "this is the object, and this is the environment in which it is located," then with any phenomenon, when the environment is not obvious, we face an insurmountable difficulty of a fundamental ontological nature.

We divide everything into entities and categories, create the metaphor "objects in a container," which guides our knowledge. But as soon as we cannot find the container, the whole structure crumbles. We are surprised to find that the objects somehow interact without touching each other, but we cannot see what connects them. This causes a model crisis. We strive to fill the gap with various abstract entities, creating the illusion of explaining such an action at a distance. We call it the Standard Model of particle physics, but it drives us away from physics to metaphysics that describes ghosts in the void.

Let's get back to the times when physicists were not engaged in magical thinking but were studying physical phenomena. Faraday went further than Chladni and placed the sand on one plate and drove the bow along the other. The figures were formed even with this interaction at a distance. Thus, he discovered acoustic induction. The analogy with electromagnetism was so evident that one step remained: to recognize that these phenomena have the same nature and begin to search for the mechanism behind them. In the case of acoustic phenomena, the medium was already known, and it was possible not to call for mysterious forces to explain such interaction of oscillations at a distance. With electromagnetism, since the medium was not always obvious, problems arose. But these problems were in the minds of theorists. Faraday, as a practical researcher who, according to some sources, made 30,000 experiments during his life, simply investigated interactions in the environment.

Faraday discovered the electromagnetic induction by winding wires on opposite sides of a metal ring and discovering the occurrence of a current in the second winding when current flows in the other, and a reversal of the direction of current flow in the second when the circuit is opened in the first. The phase change when the current was turned off was unexpected. But Faraday had a hypothesis. It had nothing to do with particles, nor with fields, nor with forces. He predicted the appearance of an "electro-tonic wave" when the circuit was closed in the first winding, which would lead to a current flow in the second. Faraday spoke of waves in a medium by analogy with hydrodynamics and acoustics.

When, after much effort, he discovered the interaction between magnetism and light (the Faraday effect), he proceeded from the wave concept of both of them. Now the effect is described as follows: a change in the polarization of light when exposed to a magnetic field. However, Faraday was not talking about fields, but about the interaction of waves. In this perspective, a change in polarization is a change in the phase portrait of the waves as they interact with each other. This is normal for waves. Polarization, as a change in the vibration vector of transverse waves, is a physical process in a real anisotropic medium: the phase velocity difference also creates a phase difference between the oscillations, which manifests itself in a change of direction. The interaction of oscillations and waves can lead to a change in the phase portrait, including an in-phase and antiphase relationship (polarization change).

Today, the Faraday effect is often associated with the Zeeman effect, which is a splitting of an atom's spectral lines in a magnetic field. Faraday made the assumption that the spectral lines can be split, but he did not have a source of powerful enough radiation to test the hypothesis. Pieter Zeeman discovered the effect in 1896. To explain the effect, Hendrick Lorenz created a model in which the atom was considered a harmonic oscillator. The equation included the angular velocity, mass and resonant frequency of the "dipole transition." Of course, in the absence of a model of the mechanism of interaction, Lorentz had to insert an auxiliary variable into the formula in the form of the "Lorentz force" with which the field acts on the charged particle. But even in this form, the equation gave solutions that led to the description of different frequency relationships with splitting into three frequencies (simple Zeeman triplet).

The planetary model of the atom says that under the influence of a magnetic field, an electron, instead of regular rotation, begins to make a complex motion relative to the direction specified by the field with a change in precession (direction of the angular momentum). This simple model made it possible to explain the observed polarization of atomic vapors' fluorescence depending on the direction of observation when the spectral lines are split into three sub levels (so-called "normal Zeeman effect"). The paradox is that the normal effect is scarce, and most often, the "abnormal" (complex) effect occurs, i.e., splitting into a significantly larger number of components with a splitting value that is a multiple of the normal splitting. When the complex effect was discovered, the billiard ball model faced a stumbling block. QM came to the rescue and explained it all by the spin-orbit interaction between a moving particle and its magnetic moment created by mechanical moment (spin). The explanation goes like this: the splitting of the spectroscopic lines of an atom occurs when the external magnetic field for an electron interacts with the magnetic field of its own spin, leading to the appearance of a "fine structure" of the electron energy spectrum. For example, the fine structure of a cooled deuterium source will be as follows:

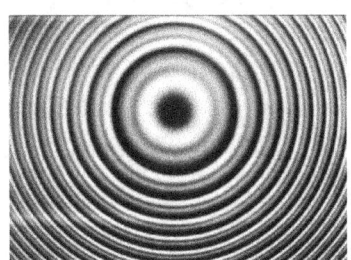

Now here comes the real paradox. All these spectral lines are discovered using an optical resonator in which a standing wave can form, and the interaction of this wave with the waves of the measured process is reflected in the form of a clear interference pattern in the device (Fabry-Perot interferometer). It is all about waves measuring waves. This effect is widely used in magnetic resonance imaging technologies. For a person not mesmerized by all those balls spinning in the void, the explanation looks trivial: it is the result of interference and spectrum splitting. Why do simple wave phenomena in a medium become flights and rotations of particles in the void? If Newton explained the effect of waves of light splitting in a prism as the flight of particles with different colors, then why shouldn't the new prophets do the same, but with the introduction of spin quantum numbers, as different "colors" of particles.

But when we think of it as a wave process, there is no paradox and both effects are absolutely normal. Suppose we perceive the observed changes in the state of the atom energy system not as a spin rotation of a ball in orbit in the void around another ball, but as a change in the phase portrait of the wave structure in the medium, and a quantum spin number as a mathematical representation of the ratios of the parameters of this structure. In that case, all splitting of spectral lines, simple (triplet) and complex (multiplet), are explainable without paradoxes. The same applies to the Faraday effect (a change in the polarization of light under the magnetic influence).

We have to return to the school desk again and remember what they usually teach when they talk about waves. Imagine a teacher who says that shear waves can only be in an elastic medium. He gives very vivid and understandable examples with a string and a drum membrane. But when it comes to electromagnetic waves, the coherent meaning breaks up. On the one hand, he is forced to say that light is a transverse wave. On the other, he is modestly silent about the fact that this wave, according to the main theories of physics, propagates in the void. If the teachers at school would go into the details of the miracles of phantom flights in the void instead of the normal propagation of waves in the environment, then the children, who had not yet had time to part with common sense, would laugh at such a teacher or at least would remain in great perplexity about the subject called "physics." Only at the university will they be forced to meet with the wonders of mainstream theories and decide whether to throw common sense into the trash-bin or not.

But suppose that some student blatantly asks about the Faraday effect: how can flying photons create the effect of a change in polarization as the direction of vibration of transverse waves? The Standard Model says: "The polarization of an electromagnetic wave is determined by a quantum mechanical property of photons called their spin. A photon has one of two possible spins: it can either spin in a right hand sense or a left hand sense about its direction of travel" (Wikipedia, "Polarization (waves)"). The teacher has to say: there is no shift in the direction of wave oscillations in the medium, but there is a change in the spin orientation of a massless ball moving in the void.

The student is confused: is this a physics lecture about the material world or a Sunday school lesson on religious beliefs? But to get rid of student's hesitation, the teacher will surely use the trance induction tool: he will begin to load the student's mind with complex formulas that describe these virtual entities and their relationships. The student can try to figure out their intricacies or just fall into a trance and memorize formulas. Subsequently, he can even learn to develop these intricate descriptions of phantoms. Such is the simple scheme of the evaporation of physical meaning from mathematical physics of the twentieth century.

But there were times when physicists needed physical meaning. Faraday described electromagnetic, acoustic, hydrodynamic and gravitational phenomena in a particular environment as a mediator of interactions. The idea of particles in the void was alien to him. He warned physicists and chemists, supporters of the atomistic billiard picture of the world, about the inevitable contradictions that arise in such a model. He foresaw that such a model would lead to the separation of theory from reality and would require the creation of "spirits" to fill the gap.

He outlined his views in a very short letter to the Royal Society, which he called "A speculation touching Electric Conductivity and the Nature of Matter" (Faraday, 1844). He wrote: "The atomic doctrine is greatly used one way or another in this, our day, for the interpretation of phenomena ... and is not so carefully distinguished from the facts, but that is often appears to him who stands in the position of student, as a statement of the facts, though it is at best an assumption ... But it is always safe and philosophic to distinguish, as much as is in our power, fact from theory; the experience of past ages is sufficient to show us the wisdom of such a course; and considering the constant tendency of the mind to rest on an assumption, and, when it answers every present purpose, to forget that it is an assumption, we ought to remember that it, in such case, becomes a prejudice, and inevitably interferes, more or less, with a clear-sighted judgment" (Ibid).

Today, any physics student will talk about the "facts" of interactions of virtual particles in the void, without even thinking that these are not facts but assumptions that became a prejudice that interferes with a clear judgment. However, the number of virtual corpuscles does not solve the main question that Faraday and others asked: what is between them? He wrote: "Hence the matter is continuous throughout, and in considering a mass of it we have not to suppose a distinction between its atoms and intervening space. The powers around the centres give these centres the properties of atoms of matter ... But if an atom be conceived to be a center of power, that which is ordinarily referred to under the term shape would now be referred to the disposition and relative intensity of forces ... The manner in which two or many centres of force may in this way combine, and afterwards, under the dominion of stronger forces, separate again, may in some degree be illustrated by the beautiful case of the conjunction of two sea waves of different velocities into one, their perfect union for a time, and final separation into the constituent waves ... It does not of course follow, from this view, that the centres shall always coincide; that will depend upon the relative disposition of powers of each atom" (Ibid).

Since by "force" Faraday understood the internal property of matter and not an external entity, it will not be a big stretch if we allow ourselves the freedom and replace the term "force" with energy. Then "the intensity and relative disposition of forces" become the frequency levels of energy and their interactions. The artificial division into matter and some intermediate space disappears, and a continuous energy oscillatory medium arises. In it, "centres of forces" are formed as combinations of the parameters of oscillations and waves, which we perceive as forms of matter.

Faraday offers an example of the interaction of real waves in a physical medium, the process of formation of structures and their decay. The hydrodynamic analogy is beautiful due to its visual clarity. But the billiard ball analogy is also quite vivid. Without a model of the mechanism of interaction, conjunction and separation, fusion and decay, the analogies can only compete in their descriptive power. At this point, it's only a matter of taste. To score the victorious goal, the "Wave" team needs a model of the mechanism. Meanwhile, the "Particle" team is building up its attack power and scoring virtual goals of hypotheses based on angel messengers flying between the balls in the void. It is not about taste but about the fact that the very essence of the scientific approach is being attacked.

At the time of the letter cited above, such an attack had not yet unfolded to its full extent. Still, Faraday knew the history of science and foresaw the danger: "My desire has been rather to bring certain facts from electrical conduction and chemical combination to bear strongly upon our views regarding the nature of atoms and matter, and so to assist in distinguishing in natural philosophy our real knowledge, i.e., the knowledge of facts and laws, from that, though it has the form of knowledge, may, from its including so much that is mere assumption, be the very reverse" (Ibid).

It is worth emphasizing the mainline of events at the beginning of the 19th century: first, it was discovered that the current in the wire deflects the magnetic needle of the compass; then they saw that between the wires with current there is interaction at a distance; then they checked whether there is a reverse effect of the magnet on electricity; it was found that the movement in the conductors generates the movement of the magnet, and the movement of the magnet generates the movement in the wire; the idea arose of separate electric and magnetic currents interacting with each other.

It is interesting to note that the induction phenomenon, though it was an obvious empirical fact, at first caused a rejection reaction among many physicists: it was not clear why "nature needed" to create current only when the magnet moved or when the current in the primary winding changed. Why is movement necessary at all? We should not forget that for a very long time, magnetism was explained by special magnetic vapors or fluids and the accumulation of elementary magnets (dipoles) until Ampere spoke of currents. Faraday was not the first to discover induction. Still, he, unlike many, was more accustomed to believe his eyes than the mainstream dogmas, and therefore was able to understand the meaning of what was happening. In any case, it became clear that this "street traffic" is two-way, but what "road" it takes remained a mystery.

At that level of practical needs, it was enough to describe at least external manifestations coherently, and the internal mechanism "under the bonnet" could wait. Everything has its time. Faraday's descriptions were qualitative (the language of formulas was alien to him), and Ampere's quantitative mathematical models were incomplete. The mathematical model of electromagnetism by William Weber, which was based on the then accepted dogmas of Newtonian mechanics that the interaction between bodies depends only on distances and speeds, generally led to physical nonsense when the kinetic energy of particles in a closed system turned out to grow to infinity. For a very long time, this did not bother anyone since the theory, in principle, described the experimental data at the required level of accuracy. The misunderstandings and contradictions of such a model could be swept under the rug for a long time by different mathematical tricks, if not for one "but": for real practice, plausible and preferably consistent with common sense theories were required. The main problem was not even infinity, but the magical instantaneous interaction, which in this case depended on the speed of the bodies. The reality was not instantaneous: a more accurate measurement revealed a delay in the interaction between charges.

Carl Gauss tried to create a theory of the mechanism of transmission of interaction but did not succeed. Bernhard Riemann described the process as a wave equation. But then, for some reason, he abandoned the idea. His article was published after his death and was not successful. The criticism's main argument was the contradiction of such a description with the Weber model, which was considered infallible. Faraday's theoretical constructions about the "lines of force" were very vague, in contrast to the brilliant specifics of his experiments.

Without a quantitative description of discoveries, the solution of technological problems would be difficult, and the tasks and prospects were so ambitious that at that moment, their entire depth and width could not be covered. Now in retrospect, we can say that a whole new era of civilization was born. Any description that allowed using the potential of electromagnetic induction in engineering solutions was an urgent need, and demand creates supply. Maxwell became the mathematician who solved this problem and laid the foundation for electrical, optical, communication and other technologies. He suggested that light is an electromagnetic phenomenon.

Interestingly, it was exclusively about the similarity of the wave properties of light and electromagnetism. Such an analogy allowed him to predict the presence of other types of radiation, invisible to the eye. These hypotheses were confirmed after 25 years in the experiments of Heinrich Hertz. He discovered invisible radiation in a specific frequency range (Hertz waves, later called radio waves). He showed that it exhibits the same wave properties as light (refraction, diffraction, polarization, the formation of standing waves). The era of wireless communications has begun.

But in theoretical physics, this led to an even more significant aggravation of the question of the propagation medium. If the phenomena of electromagnetic induction occurred at short distances, light and radio could not be called short-range. To top it all, electromagnetic radiation (including visible and invisible spectrum) could propagate in space as free from known gases as possible (vacuum). The crisis was "resolved": the environment was simply canceled, and phantom explanations of reality began.

"Maxwell introduced the "field" concept, but only as a computational device, never doubting that there was a physical mechanism operating to perform the functions involved. But starting with Einstein, and his abandonment of a substantial "aether," the "field" became a supernatural device that magically wafted energies across the void wherever needed, and allowed us to forget we had abandoned physics when we abandoned causality. Ask a physicist exactly what a "field" is made of, and how it acts to magically convey energy across the void, and you won't get answers, only hand-waving and formulae. But with this "field" devoid of any physical mechanism, a hocus-pocus wave of the wand was introduced at the heart of the discipline, and we all became magicians. Thereafter, whenever experimental evidence contradicted current theory, we had a ready-made answer. An invented "field" and its invented particles, designed to be unobservable, hence not subject to falsification, could always produce at least apparent conformity with theory. This validation of the fudge factor was then duly ratified by the Nobel Committee, in its awards for the infamous "renormalization" fudge. This meant that regardless of the experimental evidence, the theory didn't have to be modified, and no one had to change their ideas or (horrors!) learn anything new … As Carver Mead famously remarks, "It is my firm belief that the last seven decades of the twentieth century will be characterized in history as the dark ages of theoretical physics" (Hotson, 2009).

But in the 19th century, when the physical meaning was still present in physics theories, the model of the propagation of interaction could not do without the concept of a medium. Reliance on emptiness with virtual phantoms, which became the norm in a hundred years, would then sound like nonsense (which it is). Maxwell's equations spoke of the rules acting in the medium and determining the interaction of bodies, charges and currents. According to Maxwell, such a medium exists, even if there are no bodies in it. This medium is in the state of tension, and lines of force are not mathematical abstractions but the representation of this tension. He compared this tension with ropes or muscles but understood that it was only a metaphor and not an explanation of the interaction at a distance. But this does not mean that this tension cannot be used if its manifestations and regularities are described.

Maxwell replaced the concept of "force" used by Faraday with the concept of "field strength," which became the key in his model. But this was already a huge step compared to classical mechanics, which described only the interactions of bodies. He understood that the concept of the environment itself, whatever it is called, is a necessary element of the model. Still, without describing the internal mechanism, it remains an auxiliary variable, a hypothesis without filling.

Maxwell had his own hypothesis about the environment, and it was a "vortex model," i.e., oscillatory in essence. He explained the Faraday effect by the rotational nature of magnetism. He considered the medium as a combination of "molecular magnetic vortices," i.e., it was about elementary oscillations of energy. He explained the influence of the magnetic field on the light by the interaction of rotating vortices and the rotation of the polarization angle depending on the magnetic force, vortex radius, refractive index, magnetic permeability (induction capacitance), inversely proportional to the square of the wavelength. He tried to describe the theory of molecular vortices mathematically, based on the understanding that all interactions are similar in nature and mechanism.

Maxwell's model is purely mechanical: "We cannot help thinking that in every place where we find these lines of force, some physical state or action must exist in sufficient energy to produce the actual phenomena. My object in this paper is to clear the way for speculation in this direction, by investigating the mechanical results of certain states of tension and motion in a medium, and comparing these with the observed phenomena of magnetism and electricity. By pointing out the mechanical consequences of such hypotheses, I hope to be of some use to those who consider the phenomena as due to the action of a medium, but are in doubt as to the relation of this hypothesis to the experimental laws already established" (Maxwell, 1861).

He considered the vortices that constitute the medium as discrete bodies, i.e., mechanical parts of the medium. Electromagnetic interactions become the result of tension in the medium. He built a detailed mathematical model of such stress and pressure, resulting from the mechanical action between the medium's parts as "molecular vortices." In essence, it is a repetition of the Descartes model, but with a more refined description. Unfortunately, such a mechanical model of the environment, like many others, did not fit into reality, or rather, reality does not fit into it.

Maxwell's attempts to clear the way with the help of mechanical principles turned out to be insufficient even for himself. Gradually, he began to use the concept of field as a mathematical abstraction. The field concept, despite the name, which means some material environment, turned into a set of fictitious properties, designed to present some mathematical theorems describing the manifestation of physical phenomena. It was a handy tool for the needs at that time but left the question about the physical mechanism of interaction for future theorists. Unfortunately, this tradition prevailed, and after 150 years, the theoretical physics of interactions turned into a continuous set of abstractions.

Here is an excerpt from a report at one of the conferences: "The modern view of the world that emerged from Maxwell's theory is a world with two layers ... The objects on the first layer, the objects that are truly fundamental, are abstractions not directly accessible to our senses. The objects that we can feel and touch are on the second layer, and their behaviour is only determined indirectly by the equations that operate on the first layer. The two-layer structure of the world implies that the basic processes of nature are hidden from our view ... The Maxwell theory became elegant and intelligible only after the attempts to represent electromagnetic fields by means of mechanical models were abandoned. Similarly, quantum mechanics becomes elegant and intelligible only after attempts to describe it in words are abandoned. To see the beauty of the Maxwell theory it is necessary to move away from mechanical models and into the abstract world of fields. To see the beauty of quantum mechanics it is necessary to move away from verbal descriptions and into the abstract world of geometry ... In quantum mechanics just as in Maxwell theory, Nature lives in the abstract mathematical world of the first layer, but we humans live in the concrete mechanical world of the second layer ... (Maxwell's theory) ultimate importance is to be the prototype for all the great triumphs of twentieth-century physics. It is the prototype for

Einstein's theories of relativity, for quantum mechanics, for the Yang-Mills theory of generalised gauge invariance, and for the unified theory of fields and particles that is known as the Standard Model of particle physics" (Dyson, 2007).

The author, Freeman Dyson, is a professor emeritus of physics at the Institute for Advanced Studies in Princeton, USA, a theoretical physicist, one of quantum electrodynamics' founding fathers. What he calls the "great triumph" is a road that led theoretical physics to become a corpus of religious concepts. Now professors of physics preach about a two-layer world: the world of matter and the world of intangible, abstract entities. From a tool of the description of reality, math turned into a "first-layer" world of abstractions describing phantoms.

It would seem that theoretical physics, as the name suggests, should expand the boundaries of our knowledge of the material world. Unfortunately, it describes phantoms and declares reality as strange and inexplicable. It is not surprising that modern physicists wonder why the equations of theories do not work when it comes down to practice and practical physics becomes a testing ground of "miracles." Of course, if the spirits from the "first layer" are at work, then their deeds in the "second layer" are miracles by definition.

Dyson simply repeats ancient platonic idealism. Plato believed that there is a world of "ideas" as absolute and inaccessible to the direct perception entities living outside space and time. They are embodied in eidos that are accessible to our senses. Such Neo-Platonism of theoretical physicists means that the models they create are self-contained abstractions (phantom representations) that exist in some other and inaccessible world. Their ideas do not translate into eidos. The theory is not connected with the reality that it is called to model. But we must note two points in Dyson's speech from which we can draw conclusions that contradict his idealism and are of practical importance.

First, it is possible to describe phenomena at different levels of mathematical complexity. However, it does not mean that the higher we go to complex math, the more we should detach from reality. Abstractness and concreteness do not relate to the "layers of the world," but to our description level. There is no paradox in the fact that the abstract level can be simple and linear, and the concrete more complex and nonlinear. Thus, wave processes and synchronization can be described by real numbers and even linear equations. The basic laws of frequency-phase coupling can generally be reduced to ratios of integers. But as we begin to approach a more detailed description, we need nonlinear equations that do not have exact and unique solutions. But there is no paradox in this: we live in a complex, dynamic and nonlinear world of energy oscillations with a wide range of parameters.

Second, the professor notes the failure of the mechanical model. It gave useful descriptions and predictions at its level. But when it came to explaining interactions at a distance, it was useless. Moreover, all the attempts to stay within the mechanical paradigm only led to going deeper into a blind alley. It is worth noting that all the models that Dyson calls triumphs are essentially mechanical even if they speak of abstractions of the "first layer." Quantum mechanics, Yang-Mills theory, and SM are about the mechanical interactions of phantom particles.

GTR is all about the mechanical tension and curvature of a phantom space-time fabric. The first and the second cannot unite only because they created different phantoms that live in separate abstract "layers." They pretend to solve the question of interaction at a distance (all fundamental interactions are like that) but they produce only an illusion of explanation. The physical mechanism remains hidden from our view.

Does it mean that we should abandon our attempts of making physical sense of it and leave it in some unreachable "first layer"? Dyson seems to think this is the only way. But there is another one worthy of theoretical physics, if it wants to be called a science. The only way to unify all interactions in a physical model is to abandon abstract phantoms and return to reality. For this, we need to see where the previous models failed and choose another road.

For example, Maxwell used a hydrodynamic analogy for modeling electromagnetic phenomena, but exclusively in the sense of flow, pressure and tension. He built his molecular vortices model on the illustration of lines of force when scattering iron filings between two magnets. For him, the lines of force turn out to be elastic "ropes," and the attraction between the different poles becomes the connection of such ropes and their tension, and their repulsion is scattering and deflection. In other words, the lines are not a picture of the interaction of oscillations and waves, but ropes that connect the bodies.

Thus, he would have to explain Chladni figures not as the result of the oscillatory process and the interaction of waves, but as some kind of elastic "acoustic ropes." But this would be such a blatant contradiction to the reality that one could doubt the author's ability to describe physical phenomena in principle. The obviousness of vibrations and the familiarity of the medium in acoustic phenomena did not allow such a violation of common sense. But in the case of electromagnetic phenomena, the medium's non-obviousness made it possible to construct any phantom models, and they became "physical theories." In the same way as in the subsequent Bjerknes experiment, the evidence of hydrodynamic processes, the presence of a medium, oscillations, attraction and repulsion, the appearance of "lines of force" as interference pattern made it possible to draw an analogy with electrodynamics, but it again ran into an obstacle in the form of a question about the medium.

Maxwell also reduced gravity to the "rope" principle. Since he proceeded from the concept of charge type, a contradiction arose: in the magnetic effect with iron fillings, the deviation of ropes led to the repulsion of poles with the same charge, and in the case of gravity, objects with one sign of gravitational charge were attracted. Here it must be emphasized that this issue was considered and is still considered the main stumbling block for combining these interactions in one model. It is argued that any unified theory should explain the bipolarity of electromagnetism and the monopolar gravity.

The initial error in the premise gives rise to a dead-end version of the entire logical chain. However, if we get rid of the "charge type" phantom and begin to think about the interaction of oscillations with different frequency-phase coupling scenarios which lead to what we observe as attraction or repulsion, then the contradictions between different "forces" disappear as if by magic. But there is no magic: we simply free ourselves from the obsession with a hypothesis that is useful at some stage but which has become an obstacle to the development of the model in the direction of greater explanatory power for a wider range of phenomena. The mechanics of interactions could be explained by the charge type concept. Still, it contains so many internal contradictions and inconsistencies with empirical evidence that its place is in the museum of science history.

While remaining within the mechanical model framework, Maxwell tried to overcome the contradiction that this model generated. What did he propose? In the electromagnetic interaction, "the stress in the axis of a line of magnetic force is a *tension*, like that of a rope ... In order to produce the effect of attraction, the stress along the lines of gravitating force must be a *pressure*" (Maxwell, 1861). Why such identical attraction/repulsion processes in one case become tension, and in another pressure, remains beyond the scope of the model. They just should be like that, and that's it. What is their physical essence? Again, everything is limited to a reference to certain forces and their magical magnetic lines. Fixation on the mechanical model of the interaction of bodies (material points, corpuscles) leads to numerous internal contradictions. Maxwell gradually abandoned attempts at physical modeling and went into mathematical abstraction called "field" with a set of fictitious, but very convenient properties for solving equations.

In his article in the section of the British Encyclopedia of the 1878 edition entitled "Ether," Maxwell wrote: "If there is any motion of rotation, it must be a rotation of very small portions of the medium each about its own axis, so that the medium must be broken up into a number of molecular vortices" (Maxwell, 1878). Words are critical because they have to carry meaning. Maxwell gave the word "vortice" a mechanical meaning and used an analogy with a wheel. However, there is a hydrodynamic and aerodynamic meaning, which leads to an analogy with water whirlpools and vortices in the atmosphere. They can rotate in different directions, interact with each other, attract, repel, establish more or less balanced relations. They can disappear, absorb each other, and generate new ones. All this happens in a continuous environment.

To see such processes, we don't even need to be in the center of a hurricane. It is enough to go on a nice sunny day to the riverbank and sit there quietly, watching what is happening. Whirlpools with different sizes, speeds and directions of rotation continuously form in the water. They can remain independent, but they can also merge. They arise and disappear. They repel and attract. The dynamics of the interaction of oscillations with different amplitude-frequency characteristics and phases in a single continuum is evident. In five minutes of observation, we can see the macro- and microcosm model in action.

Now imagine that we do not see water, but only whirlpools. Furthermore, we do not see many whirlpools because of their small size, but we can observe some external manifestations of their interaction. Please note that in the water both whirlpools and "lines of force" (interference patterns) form. The disappearing whirlpools leave traces. New ones arise and continue the process. They interact or not depending on their parameters, distance, and the state of the medium. But we see only large whirlpools and some traces of small ones. The rest and the medium itself remain beyond the resolving powers of our natural perception and even artificial devices. Is it possible to infer the laws of a process by simply observing its phenomena? Why not? But we need to include in the model the environment where these phenomena happen even if we cannot measure it directly.

The example of whirlpools in the river was intended to demonstrate the futility of modeling the process as the rotation of individual particles in the void. But it was also limited because vortex rotations are not the only form of the oscillatory process. Such a model of macrocosm and microcosm, with all its clarity, will be a simplification leading to errors. A rotation is an oscillation, but not every oscillation is a rotation of an object. Energy vibrations can manifest themselves as the rotation of individual bodies, but this is an epiphenomenon (secondary manifestation) of internal oscillations and their interaction with external ones. If we fixate on this narrow aspect of energy vibrational processes, we begin to see rotation of the particles even where there are no particles or rotations.

Rotations of objects obey the same laws as other forms of oscillatory processes. We can demonstrate this with another curious thing from the exhibition. There is such an interesting toy called "rattleback":

This is a kind of spinning top. The difference between it and the usual one is that it can change the direction of its rotation independently. When turning it in one direction, it first rotates in a given direction, then it starts to decrease angular velocity and increase the amplitude of oscillations, and then it sharply changes the direction of rotation. Some design options may exhibit this behavior with a spin in a specific direction, while others are independent of the original direction. You can make such a spinning top from a spoon, and it will demonstrate several changes in the direction of the spin (see the video in the Wikipedia article "Rattleback").

The design of such a toy appeared empirically. It can be described geometrically. We can even model the distribution of mass relative to the axis of symmetry. It usually has a shape in which the lower surface is an ellipsoid segment and the upper is a plane. The axis of symmetry of the mass has a slight rotation in the horizontal plane relative to the axis of symmetry of the supporting surface of the ellipsoid.

Having given such a mechanical description, one might think that the phenomenon is explained and be satisfied with it. But this is not an explanation. It is merely a statement of fact: a structure behaves this way because it is so constructed. But the question arises: why, with just such a design, spontaneous changes in the direction of rotation occur? Why doesn't it behave like an ordinary spinning top that obeys your will and spins only where you turned it? One may ask a question like this: why does this oscillatory system demonstrate a change in the phase portrait passing through the bifurcation points? Why is the phase portrait unstable with respect to the direction parameter?

It has instabilities due to asymmetry in the distribution of mass along the axes of rotation and swing perpendicular to each other. These parameters are in certain proportions, and depending on them, the rotation of the top has a specific phase portrait. So, some designs show a change of direction when they begin to rotate in the direction of instability of the swing axis. In comparison, others change direction several times, regardless of the original direction. The rotation of such a top can be caused by giving it angular velocity and clicking on one of its ends or swinging it. Thus, we observe a complex nonlinear dynamic oscillator system with many parameters. A detailed study of different modeling options for driving external forces, friction, and resistance showed that oscillations of such a system on a harmonically oscillating base could lead to periodic, quasiperiodic, and chaotic phase portraits with rich bifurcation dynamics (Awrejcewicz, Kudra, 2014).

We see that the phenomenon can be described as the mechanics of the material point (body, corpuscle) movement and as a complex oscillatory system interacting with the environment. For both cases, different models can be used, and the parameters can be named differently. The mechanical model itself is neither bad nor good. We cannot say that it is entirely wrong. It has its own truth. But it is limited, and this limitation is especially evident when the question of interactions arises. It can describe interactions as the relations of discrete material points and some forces acting on them. This description is linear, mechanistic in the worst sense of the word: tearing the whole apart, it tries to create a model for the interaction of these parts. Where there was continuity, the integrity of the process, it creates a picture of individual "gears" of the mechanism, between which there is the void. It is not surprising that it encounters contradictions when it is necessary to explain the complex nonlinear dynamics and the interaction of such "gears" without direct contact (action at a distance problem).

As we noted in the first volume of the study, a continuous energy signal can only be measured discretely by sampling and represented as a sequence of samples x[n], where n is an integer (quantization). Analysis of any signal is its decomposition into discrete measurements, and synthesis is their combination. The more frequently the sampling is made (higher sampling rate), the better the synthesis and the closer to the original signal continuity. There is no point in the temporal dimension that can be called the limit of limits so that the combination of such points can be completely equated to the original continuous signal. We can only make the representation smoother and closer to the original continuity. But even at a very high frequency, each measurement will be a quantum (part of the original signal).

Even a very short signal is a continuous process. If our sampling frequency coincides with the signal duration, then it will turn out to be a "point" delta signal (Dirac delta function). It will have a specific value at the measurement point but zero at the other measurement points:

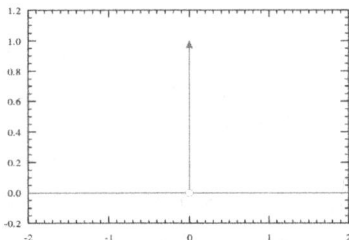

We can fall into the objectification trap and think that we have found a particle as a discrete material point. But such a delta signal, having a value of 1 at the time of measurement and zero values on the left and right, is a mathematical abstraction. Real signals are continuous and always "blurred in time" (they have dynamics of change). If a new signal has arisen, then it will still be a continuation and transformation of the previous energy flow. If the signal disappears abruptly or exponentially dies away, then this is only a transition of energy from one state to another. What is commonly called energy dissipation is a phase of the general energy process in which scattering transforms into concentration and vice versa. The entire energy continuum of the Universe can be represented as the dynamics of changes and transitions of energy levels. If these were ideal monochrome signals in an ideal isotropic medium, everything could be described as continuous sinusoids passing into each other and superimposed. The real picture is more complicated since there are neither ideal monochromatic waves nor ideal isotropy. However, the picture's complexity does not mean the impossibility of its description and modeling.

Discretization and integration are two sides of the same "coin" of cognition. Without sampling, integration is not possible, but without integration, sampling does not make sense. The bias towards analysis to the detriment of synthesis is a pathology of the state of the reality modeling process. As the name suggests, Quantum Mechanics should explain the mechanism behind the integer ratios of quantum measurements of energy processes that have been observed since the end of the 19th century. But the version that is considered an "explanation" is not physical but metaphysical: it postulates the mystical quantum properties of particles that appear to be integers without any specific reason. This is the outcome of the bias towards the discretization and objectifying the discrete measurements as material points (particles). The rest of the errors follow from these basic ones.

The Theory of Energy Harmony, which offers the synchronization (frequency-phase coupling) as a mechanism of interaction of different energy levels with different frequency characteristics and phases, not only explains the integer ratios as harmonics but also shows how stable structures can be created and how they can decay with the release of energy in the same integer (quantum) form as when it was absorbed. It does not need to call for the magical birth of particles from nowhere or annihilation into nowhere. It does not contradict the law of energy conservation. It does not need to breach the physical causality by saying that particles travel back in time. It just speaks of oscillations and waves in the universal energy environment which interact with certain patterns that we can measure discretely and get the integer values when they tend to stable harmonic ratios.

The problem of the corpuscular "billiard ball" model of the world is not even in the concept of material points as elementary parts of matter. After all, balls can interact in many ways. But even such a mechanical picture needs a medium of interaction. Billiard balls come into direct contact but it would be impossible if not for a table they are placed on. Billiard ball analogy does not do justice to the QM or SM as they speak of magical interactions of "balls" in the void. Even the concept of the field does not help as without an interaction mechanism description it turns into a math abstraction. The "balls" hang in the void, and no description of their flights' geometry will help. We can draw diagrams of trajectories ad infinitum, but without the physical medium, their movements will seem magical and unpredictable to us.

If we consider the world as a single energy oscillatory system, then the phenomena become manifestations of the processes occurring in it and creating structures that are stable for some time. We can call them anything we like. The basic understanding that the interaction of these structures at a distance or in direct contact is determined by the mechanism of oscillatory processes in the entire system does not allow us to get into the blind alley, even if some level or element of the system is not available to us for direct measurement.

The authors of the book "Synchronization: A Universal Concept in Nonlinear Sciences" wrote: "In many cases the oscillators cannot be considered as discrete units but rather as a continuous oscillatory medium" (Pikovsky, Rosenblum, Kurths, 2001). The authors summarize the accumulated knowledge in many fields about oscillatory processes and their synchronization, give general intuitive descriptions and detailed mathematical models, present diverse examples in inanimate and living nature, repeatedly emphasize the universality of phenomena, and say that their book is perhaps the first attempt at an interdisciplinary approach. But they bypass the macrocosmic and atomic levels of matter and the problem of fundamental interactions as if it were a taboo.

The Theory of Energy Harmony "stands on the shoulders" of all the researchers of sync starting from Christiaan Huygens in the 17th century to the scientists of modern days. It adheres to the only taboo: the ban on any taboo. It follows the main principle of scientific research: "There is no place for dogma in science. The scientist is free, and must be free to ask any question, to doubt any assertion, to seek for any evidence, to correct any errors … We know that the only way to avoid error is to detect it and that the only way to detect it is to be free to inquire. And we know that as long as men are free to ask what they must, free to say what they think, free to think what they will, freedom can never be lost, and science can never regress" (Oppenheimer, 1949).

The dogma of the corpuscular model of the world has to be put on the shelves of the history of science. We detected this error and corrected it thanks to our freedom to doubt any assertion and ask any questions. We sought and collected evidence that speaks of the continuous nature of the Universe at all levels. Thus, we can reformulate the above statement of the authors: to explain the mechanism of fundamental interactions, we should not consider the oscillators as discrete units but as elements of a continuous oscillatory medium.

Here is how the authors describe such a medium: it has homogeneous profile of the phase with synchronization between all points and constant coupling of oscillators with a distance between them tending to zero; the interaction of oscillations forms complex structures that can be sources and receivers of waves; the velocity of their interaction propagation depends on the inverse square of the distance; the frequency profile has local maximums and maximums of the potential depending on the sign of the coupling coefficient; a local inhomogeneity having a larger/smaller frequency than the environment emits waves and gives rise to patterns of interaction (Pikovsky, Rosenblum, Kurths, 2001).

The authors do not make any conclusions that could be associated with the problem of fundamental interactions and their medium. As we have noted, they avoid this area of theoretical physics as if it is a restricted area that only the phantoms of the mainstream models can visit freely. But suppose we disregard the prohibitions and remember the freedom of thought principle. In that case, we can see that such a description applies to an all-encompassing energy environment (call it ether, field, or anything). Such an environment will demonstrate the combination of qualities that confounded all theorists of mechanical models of the ether. When we consider its properties from the continuous vibrational processes and their synchronization perspective, they begin to acquire physical outlines. The description of interactions in such a medium can be transferred from the metaphysics of the mainstream theories with their phantoms in the void to the field of physics.

If we proceed from the description of the medium to the description of the frequency-phase coupling mechanism that works in such a medium, it will turn into an explanation of fundamental interactions and all of the observed phenomena of attraction-repulsion and balance, creation and destruction of material structures, fusion and decay, emergence and propagation of waves of interaction. Last but not least, we can explain why all these processes in all kinds of interactions demonstrate universal dependence upon amplitude-frequency parameters and phase portraits of interacting structures and the medium. The idea of the physical mechanism is the "Ariadne's thread" that helps us get out of the maze of theoretical physics muddle.

CHAPTER 8

THE HARMONIZATION OF CHAOS

Perhaps the need to find or feel some ultimate harmony or order is a universal of the mind.

Oliver Sacks

The book written by mathematician Steven Strogatz is called "Sync: the emerging science of spontaneous order" (Strogatz, 2003). The title contains an important message. Perhaps, by studying synchronization, we will find the answer to the basic ontological question of how order arises from chaos. It is an emerging science, and it does not even have a name. The word "sync" means a phenomenon, but the field of knowledge about it needs a name. We can call it *Synchrology* (understanding synchronization).

Here is how Strogatz ends his book: "For reasons I wish I understood, the spectacle of sync strikes a chord in us, somewhere deep in our souls. It's a wonderful and terrifying thing. Unlike many other phenomena, the witnessing of it touches people at a primal level. Maybe we instinctively realize that if we ever find the source of spontaneous order, we will have discovered the secret of the universe" (Ibid).

In one phrase, Strogatz expresses the balance of basic emotions: "wonderful" (causing curiosity) and "terrifying" (causing fear). An individual can succumb to fear and stop a personal search, but on the whole, humanity as a species is bound to search for answers. It is the essence of the process of creating an adaptive model of reality.

Just one seeker is enough for the search to exist. But there are many seekers trying to answer general questions in line with specific tasks and their solutions to reveal the secrets of the Universe. However, specific questions cannot be answered if there is no general model. Questions must be asked from a certain perspective. Halfway to success is the correct formulation of the problem. If there is no concrete hypothesis regarding the answer to a general question, data "hangs in the air" without support for all its empirical validity.

Data always test an assumption, even if it seems that it is a "pure" collection of information. Data may turn out to be unexpected. We can find what we were not looking for. But this does not mean that there was no search for something and no original idea. To find a black cat in a dark room, we need to have a cat model. We must know what we are looking for; otherwise, the process is even more meaningless than looking for a nonexistent cat.

Strogatz describes how numerous explanations for firefly synchronization have been put forward over the decades. Initially, there were two extreme versions: the fireflies have a supervisor or conductor, or all this is an accident. For a long time, it was overlooked that the fireflies did not flicker at once and all in unison but do it in a certain rhythm. Even alone, they can hold a basic pulsation regulated by neural rhythms (internal synchronization). The only thing left is to synchronize with external signals. There is no conductor. It's not a random chance, either. It is a process of oscillators self-organization, in which there are patterns of phase and frequency interactions. Main conditions: the presence of self-sustaining oscillations and communication between oscillatory systems. Of course, fireflies do this for the purpose, but the universality of the physical phenomenon means that it exists and works on its own, and living systems simply use it.

As Strogatz correctly observes, such a vivid phenomenon as the synchronization of fireflies has highlighted (in the literal sense of the word) the essence of many puzzles of modern science and shown a new facet of the question of self-organization of complex dynamical systems from atoms and cells in living systems to galaxies and the Universe as a whole.

But not all oscillators synchronize and not always. Relations are established and broken. What are the conditions for synchronization? What are the determining factors? The need for synchronization model arose. Help came from the biologist Arthur Winfree. In his model, oscillators have an influence function and a sensitivity function (Winfree, 1980). If we use the relay race metaphor, we can compare the influence function with how one runner tells the other "run faster, run slower" (depending on the separation of their phases). But the runner not only speaks but also listens. The sensitivity function determines the way he listens.

Function values are inserted into the formula. If the influence function has a positive value, then one has to run faster; the negative — slower, zero — no change. A positive sensitivity function means that the runner perceives the signal adequately, negative — he got it wrong. The Winfree model implied the possibility of acceleration and deceleration depending on the signal, its characteristics, on the phase point. This theoretical and mathematical model coincided with what was observed in real experiments with biological oscillators. He averaged the function of influence and sensitivity, but various frequency characteristics of the oscillators were embedded in the model. Runners could run at different speeds.

But the question arose of who was giving the command to slow down or accelerate. If we take the example of fireflies, we can assume that everyone is oriented towards a neighbor. But a complex system is not linear, and the neighbor may not be the one who lives on the same "floor." Imagine that a neighbor is a friend in a social network, but he is on the other side of the globe. The question arises of organizing such a network for effective and quick communication.

Winfree decided to omit this problem because the model was very complicated anyway. His equations described the phase space of the oscillators, which were influenced by three factors: speed (proportional to the natural frequency of the oscillator), current sensitivity to influence (dependent on the phase point), and the general effect of all other oscillators. The differential equations system made it possible to describe the current state of the system and make a prediction about future states. It is the goal of any model.

Winfree began to simulate the behavior of the model as parameters change. He found out that under specific parameters, the population of oscillators spontaneously synchronized: at first clusters formed, then more and more participants joined them. It turned out that there was no linear hierarchy. No one was a conductor or an obedient performer. The frequency of synchronization did not necessarily coincide with the natural frequency of any of the participants. It was a population of related clusters or families. It turned out that extremely heterogeneous populations did not synchronize. As soon as the critical threshold of diversity was overcome, synchronization took place spontaneously, and then more and more "runners" joined the general relay race.

It was like a phase transition during the freezing of water: a new state arises from the old through bifurcation. These are not discrete states separated by an abyss, but a continuous process. The width of the frequency distribution is similar to temperature: at some point, the temperature passes the bifurcation point, and crystals begin to form. If we recall that temperature is a measurement of energy, vibration frequency, then the analogy is not surprising. Winfree conceptually introduced synchronization as "crystallization in time."

Physicist Yoshiki Kuramoto became interested in Winfree's work, took one of his model versions, and showed how it could be solved. It was a system of a vast number of nonlinear differential equations. Such systems are rarely solvable. What did he do? He replaced the variables of "influence" and "sensitivity" with a symmetric rule of interaction with frequency synchronization. This was called "coupling strength." It determines the ability of oscillators to adjust their frequency for synchronization.

There is no need for specific points in the phase space where everything could meet. The relay baton does not have to be passed all the time in one place. As runners move, they just have to adjust speed (frequency) depending on the relative positions. An external milestone is not required, as oscillators become reference points for each other. The phase and frequency spacing are essential. Even if there are many runners, they can orient themselves to the average sync field as the common frequency.

To determine the degree of synchronization, Kuramoto introduced the concept of "order parameter." Its value is always between 0 and 1 and is calculated by the formula, where the relative positions of all the "runners" are included. Complete unison is 1, chaos is 0 (see the animation of the model in the Wikipedia "Kuramoto Model"). Kuramoto suggested that over time, this parameter for a given system becomes stable. Even 0 (no synchronization) is a possible outcome. There is also an intermediate option when there is a synchronization cluster and not joining "renegades." It depends on the width of the frequency distribution. At a particular threshold value, synchronization is not achieved because there is too much variety. This is of great importance for the analysis of the system of oscillators. The presence of similar characteristics and clusters with a narrow distribution range of these characteristics determines the synchronization order.

The Kuramoto model was a system of equations that described the state of the system with great accuracy and with a limited set of variables. It almost answered the age-old question: how does order emerge from chaos? Strogatz decided to pose this question using incoherence as a measure of entropy and equilibrium as entropy. Equilibrium is simultaneously the highest entropy and the most stable state. For example, a cup of cooled tea, which has reached maximum entropy, is also stably cool, because it can't warm itself up. But there is another concept of equilibrium in the sense of balance. This is an unstable state, in the sense that the opposing vectors can shift it in their direction. The question arose: is synchronization a stable equilibrium or unstable? Instability, in this case, means an opportunity for self-organization, for spontaneous change of state. Intuition tells us that synchronization is an unstable equilibrium. But the problem is that chaos is endlessly diverse. How can something coherent form out of an infinite set of incoherent states? How can it remain in unstable equilibrium and not go over into the entropy of a complete and stable equilibrium as the dispersion and decay of a system?

Strogatz recalls that half asleep, he saw the image of a multi-colored liquid, where each color meant the natural frequency of the oscillator. These colors do not change because the oscillator does not change its fundamental frequency. It was an image of an incoherent state. It remained to understand what effects would lead to synchronization. It turned out that if the oscillators are very close in characteristics, then the effect grows exponentially and phase coupling occurs. However, the Kuramoto model said that stability did not depend on frequency parameters. Strogatz writes that they were at a dead-end until a colleague told them that physicists from another field had known this phenomenon for 45 years. It is called Landau damping and speaks of waves in a plasma. This attenuation occurs in superfluid helium II (discovered by Peter Kapitsa in 1938) and is explained by the interaction of resonances in the medium with electromagnetic waves arising in the plasma. These resonant elements interact, and the wave energy decreases. It all depends on the distribution of frequency ranges. If there are more high-frequency wave packets ("fast electrons"), they give energy to the wave; if there are more slow ones, the wave loses it. It was about the fundamental laws of the propagation of waves in the physical medium, which we have already considered.

Strogatz notes that no one has previously linked the phenomena in superfluid plasmas, mainly inhabiting the Sun or other stars, with the pulsation of fireflies by the river. The players are different, but interaction patterns are the same. He applied Landau's calculation techniques to the Kuramoto model, and a lot fell into place. But this was a simplified model. The model considered the oscillators as close to identical, and the interaction between them as a sinusoidal dependence on the phase difference of each pair of oscillators. How can one measure the interaction of a couple of oscillators in a system if a complex system differs in that everyone influences each other? The selection of a pair from the system (physical, logical, mathematical) disrupts the whole process.

Let's return to the Huygens' experiment from which it all started. When he observed the interaction of two clock pendulums on the wall it seemed like a miracle to him and he called it a sympathy. But when he hung the clocks on the beam between the chairs, he understood that the secret was the connections between system elements. It was not just two pendulums miraculously sympathizing with each other at a distance. To sync the oscillators should be part of a system with certain relationships. It is not enough to describe the system's elements, their characteristics as oscillators, their frequency, and phase relationships. We need a description of the entire system, and then the synchronization will no longer look like a miracle.

Even several hundred years after Huygens' experiments, many perceive pendulums' synchronization as an accident or a trick. It is neither an accident nor a trick. Huygens discovered one of the leading, if not leading, interaction in nature. What does it depend on? After all, this is not an absolute transcendent entity that defines itself. Although, if we fall into the error of objectification, we can come up with a new god, and his name will be Synchronization. But this process is immanent to matter. It depends on the interacting elements themselves, their separate characteristics, and the characteristics of the system as a whole. Synchronization is an interaction, and that says it all.

We can use our mind and purposeful will, which itself is a manifestation of synchronization, to create synchronization phenomena. We can use synchronization in power grids to transmit energy. We can synchronize electromagnetic waves in a laser beam that carries waves of the same range in one direction in one phase to transmit and process energy-information. We can create synchronization of elements in computers and their networks for the transmission and processing of energy-information. We can synchronize instruments in an orchestra playing a symphony. All these phenomena created by our mind are based on synchronization. It means one thing: we do not invent but discover and use this method of efficient energy transfer existing in the Universe.

For example, the interconnection of power grids and the alignment of the frequency standard provided tremendous benefits in the efficiency of production, transmission, storage and use of energy. But without generator synchronization, this would not be possible. Now, if suddenly any generator falls out of the synchronized ensemble, special protection forcibly disables it. But what is surprising is not that it was created, but that it created itself. Engineers found out that the generators connected in the network synchronized spontaneously. They began to "sympathize" with each other like Huygens pendulums. The network of generators is a self-organizing system of oscillators, where the order is born spontaneously from chaos, where coherence is the basis of the system's existence. Violation of coherence, desynchronization means a pathology of the system, the breakdown of bonds. What's the secret? If the oscillators are connected in a network with feedback connections, they begin to affect each other's frequency and phase characteristics. The fast ones give off energy and slow down; the slow feed on and accelerate.

Synchronization is used in all communication systems: from the good old radio to cellular and satellite. Synchronization is the basis of global banking settlement systems and global positioning systems. They require tremendous accuracy, and a split second out of sync means errors that bring the whole system to nothing. Huygens pendulums could synchronize, but an increase in the distance by several meters led to a disruption in synchronization due to the deterioration of communication in the system. Now we are synchronizing the oscillators at different ends of the planet and even synchronizing the equipment on Earth with the equipment on a distant space probe. If we have achieved such successes with virtually no unified theory of synchronization, then what is the prospect if we can bring our disparate knowledge into a coherent concept?

At the moment, the theory of oscillations and synchronization has a modest role of Cinderella. But we remember that in the fairy tale, it was Cinderella's foot that fit the glass slipper, and she became the princess. Her half-sisters, who had long imagined themselves to be the main ones, remained aloof. However, miracles do not happen by themselves. You have to work, separate the wheat from the chaff, and then you can go to the ball. Strogatz wrote that perhaps a zeitgeist has come. At the end of the 20th century, articles and books on the synchronization of oscillators began to appear in various fields, from neurobiology to geophysics. People began to draw analogies that had never occurred to them before. What seemed completely different showed mutual patterns. Strogatz wrote: "Although it was always clear that groups of living and nonliving oscillators were each prone to synchronize spontaneously, it was only in 1996 that we realized how similar the underlying mechanisms can be. The resemblance, it turns out, is familial — a sign of the same mathematical blood" (Strogatz, 2003).

For example, he describes synchronization occurring in arrays of Josephson junctions. We have dealt with this superconducting effect and its explanation from the TEH perspective in the previous chapters. Here we will stress that classical electrodynamics has not explained the effect. Moreover, the phenomenon contradicts the basic laws of this model. Alternatively, SM traditionally tries to "explain" the superconductivity phenomenon by postulating new rules for virtual particles' behavior. But even this does not help as the reality comes up with surprises for a model that does not have an idea about the actual physical mechanism. It is a typical story of a bad model bluffing for the sake of staying in the game. When Brian Josephson first predicted such a phenomenon in 1962, one of the creators of the Bardeen-Cooper-Schrieffer (BCS) theory of superconductivity that explained it with virtual phonons, John Bardeen wrote in a review that it just could not happen. When it happened, the only thing that SM theorists could come up with was this: "It is an example of a macroscopic quantum phenomenon, where the effects of quantum mechanics are observable at ordinary, rather than atomic, scale" (Wikipedia "Josephson effect").

What are these macroscopic quantum phenomena? They are "processes showing quantum behavior at the macroscopic scale, rather than at the atomic scale where quantum effects are prevalent. The best-known examples of macroscopic quantum phenomena are superfluidity and superconductivity ... Macroscopic quantum phenomena can be observed in superfluid helium and in superconductors, but also in dilute quantum gases, dressed photons such as polaritons and in laser light. Although these media are very different, they are all similar in that they show macroscopic quantum behavior, and in this respect they all can be referred to as quantum fluids" (Wikipedia "Macroscopic quantum phenomena").

If not for a reference to "dressed" phantoms, we see that all of the above examples talk about physical media where physical wave interactions happen. Why should we call normal synchronized coherent wave structures such as lasers or semiconductor optical resonators "quantum fluids" would be a mystery for us, if we did not know the tradition of the "Particle" team to name all wave phenomena that show quantized (discrete) ratios as the interaction of virtual quanta-particles. For example, the above-mentioned polaritons are defined as quasiparticles resulting from the strong coupling of electromagnetic waves with an electric or magnetic dipole-carrying excitation. Why coupling of oscillations and waves is defined as particles? No wonder: traditional corpuscular dogma is based on the objectification error.

But for a person who looks at reality without the blinkers of dogma, all of the above effects, including the Josephson junction, are just confirmations of the same "mathematical blood" that describes the underlying physical mechanism. When Strogatz and his colleagues studied this effect, they realized that there was a model of coupled oscillations that well described the patterns of the observed phenomenon. "The Kuramoto model has always been a solution waiting for a problem. It was never intended as a literal description of anything, only as an idealized model for exploring the birth of spontaneous order in its simplest form. Yet its newfound connection to Josephson arrays immediately explained why these devices should synchronize abruptly. The phase transition was fundamentally the same as the one that Winfree had discovered in his model of biological oscillators, and that Kuramoto had later formalized so elegantly in his solvable model. Experts on Josephson junctions had seen this transition in their own computer simulations, years earlier, but without a theoretical basis for understanding it, it had never attracted attention (illustrating the adage that you should never trust a fact until it's been confirmed by theory)" (Ibid). The paradox of this humorous adage is that it is not a joke. If there is no model to explain the fact, it is often ignored.

Strogatz gives an example where ignoring sync led to an engineering fiasco: the Millennium Bridge in London. This bridge was to become a symbol and pride of the city but became a shame. The design was amazing (it was called the "blade of light"). The Queen solemnly opened the bridge, people entered, and the bridge began to dance a wild and uncontrollable dance, like a wriggling snake. No computer simulations have predicted this, although the phenomenon of bridges' resonances is known and seems to be studied. Two days later, the bridge was closed. Then they realized that pedestrians synchronized not only in movement along the bridge but also across. A person while walking deviates to the sides. The frequency of this oscillation ($\approx$ 1 cycle per second) is half the frequency of the longitudinal oscillation. This frequency turned out to be resonant for the bridge. The bridge's oscillations forced people to balance, which led to an increase in the amplitude of the resonance. This was the emergence of spontaneous order from the chaos of multiple vibrations. By the way, even before the months-long experiments showed the reason, two days after the bridge was closed, Josephson wrote an article in the newspaper, where he explained the problem. He understood what synchronization was.

Oscillators can be different, but there is one synchronization mechanism. When we see this with striking examples, there is a feeling of delight. Look at two completely different at first glance phenomena: the synchronization of a large number of metronomes and the synchronization of thousands of birds in a flock. We already spoke about metronomes, and now the turn of birds has come. This amazing synchronization show of a huge number of birds got scientists interested. They studied the phenomenon with the help of high-resolution cameras and computer simulations.

Previously, it was thought that one bird made the decision, and the rest followed it. This was a classical linear approach and a search for a conductor in the ensemble as a guiding and directing force. The spontaneous emergence of order from chaos was not considered. But with so many participants in the process, everyone can't follow one leader immediately, and the dissipation of energy-information would not allow the whole flock to act synchronously. A detailed analysis showed that some birds form high-density clusters, synchronize, and trigger the overall process. But such a leading role does not belong to any single team. There is a spontaneous change in the composition of the clusters, a democratic rotation of power. The primary synchronizing role always belongs to some dense group. The signal propagates from it at a speed of 20-40 meters per second. This means that a flock of 400 birds needs only half a second to synchronize the turn.

Researchers Andrea Cavagna and Asja Jelic from the Institute of Complex Systems in Rome made a mathematical model based on observations of starlings' flocks (Cavagna, Jelic, 2014). They found out that each bird has a specific characteristic as an element in a complex system. If we look at each bird as a "quantum" of the flock, it becomes clear that when we select it as a discrete part, the meaning of the whole process eludes. If we are talking about the formation of a structure and the interaction of its elements, then the characteristics of one bird acquire meaning only in relation to others. Moreover, the process of matching frequencies and phases of elements (starlings) is impossible without an intermediary (in this case, an air) that provides not only for the oscillatory process of the elements themselves but also for the exchange of energy-information between them.

It was difficult for starling researchers to invent any special "quantum" properties of virtual entities, since this level of matter is quite accessible to direct study. They had to measure the real parameters of the process: the ratio of frequencies and phases that allow starlings to synchronize and interact with the tremendous speed of energy-information transfer. They analyzed the "mathematical blood" of the process. From the point of view of the Theory of Energy Harmony (TEH), the result is predictable. The algorithm for the occurrence of synchronization in a flock of starlings mathematically resembles the behavior of superfluid helium in a coherent state at temperatures close to absolute zero. Coincidence? Accident? Miracle? Yes, a coincidence, but not an accident. It is a regularity arising from the coincidence of the laws of the joint mechanism. Yes, it is the wonderful phenomena of the world, but not a miraculous wonder. It is a physical reality that can and should be modeled without involving phantom auxiliary variables in the form of virtual transcendent entities.

A new look at familiar things led to the fact that some "physicists took time off from pondering quarks and black holes and began to pay attention to more mundane phenomena that they'd previously dismissed as annoyances: the fitful pulsations of unstable laser beams, the noisy voltage oscillations of certain electrical circuits, even the dripping of leaky faucets" (Strogatz, 2003).

Not everyone can deal with the persistent creation of phantoms in the form of virtual quarks and no less virtual black holes. Of course, the mainstream continued to be where funding and bonuses were, but many tried to explore real uncharted territory instead of producing imaginary maps of virtual lands. Even ordinary drops from the faucet turned out to be symbols of nonlinearity (chaos in this sense) and could not be described by linear equations. A linear system can be divided into parts, solved separately, and connected again. But often, it is a very approximate model. You can make an engineering calculation, but it will work only in a specific range and in some circumstances. All unsolved problems of sciences from physics and economics to biology and psychology concern nonlinearity.

Arkady Pikovsky and Mikhail Rosenblum, in their article "Synchronization: from pendulum clocks to chaotic lasers and chemical oscillators," wrote: "It is important to emphasize that synchronization is an essentially nonlinear effect. In contrast to many classical physical problems, where consideration of nonlinearity gives a correction to a linear theory, here the account of nonlinearity is crucial: the phenomenon occurs only in the so-called self-sustained systems ... The each-to-each interaction is also denoted as a mean field coupling. Indeed, each firefly is influenced by the light field that is created by the whole population. Similarly, each applauding person hears the sound that is produced by all the other people in the hall. Thus, we can say that all elements are exposed to a common force. This force results from the summation of outputs of all elements ... Thus, the oscillators in a globally coupled ensemble are driven by a common force. Clearly, this force can entrain many oscillators if their frequencies are close. The problem is that this force (the mean field) is not predetermined, but arises from interaction within the ensemble. This force determines whether the systems synchronize, but it itself depends on their oscillation — it is a typical example of self-organization" (Pikovsky, Rosenblum, 2003).

It is an example of cyclic causality. There is no external force organizing the process. Synchronization is not a magical entity but a process of interaction of energy oscillations. Although the terms "mean field" and "coupling force" evokes associations with the regular use of the words "field" and "force" in physics, there is a clear initial position in the theory of chaos and nonlinear systems: it is not an entity transcendental for a system, but an internal process, the dynamics of which show how synchronized the system is and whether it is a system of interacting elements. The mean field, as a variable in the equations describing ensembles of synchronized oscillators, does not explain the process but itself requires an explanation and description of its dynamics depending on the processes in the observed and simulated system.

In such a conceptual base, there is an understanding of the nonlinearity of processes and auxiliary variables in the form of entities that can explain everything by themselves, including themselves, are not used. On the one hand, it puts the theory of dynamical systems in a difficult position: it requires rigorous scientific modeling. One cannot say that some entity makes everything happen as it happens, but its ways are inscrutable. On the other hand, it became a vaccine against the errors of objectification and the temptations of moving away from science into mysticism with transcendent entities of any kind.

The term "chaos theory" emerged as a response to observed nonlinear behavior in complex dynamical systems. Unlike simple periodic dynamics, it gave the impression of chaos without regularities. But the name turned out to be misleading about the essence of the theory. It is not a rare occurrence in science. For example, the Theory of Relativity actually speaks about absolutes. Chaos theory speaks of spontaneous, self-organizing and self-sustaining order. It describes how different systems can exhibit the same patterns of emergence of order and transition to chaotic behavior.

It seems that chaos and harmony, being opposites, have no connection. But chaos can synchronize, and harmony can turn into noise. They do not "sit" on the opposite sides of the swing, as they are not separate entities. It is the process of energy transformations and structural changes. Chaos and harmony are just two sides of the same oscillating "coin" of energy that make mutual transitions. These transitions have specific patterns that can be analyzed and modeled.

When researchers realized that synchronization did not bypass complex aperiodic systems, it radically changed the understanding of the process. Synchronization was usually associated with a strict periodicity. But it is not limited to simple rhythms and melodies. Polyrhythm and polyphony can be synchronous too. In music, polyrhythm is a simultaneous use of two or more rhythms that are not readily perceived as deriving from one another or as simple manifestations of the same meter. Polyphony is a musical texture consisting of many simultaneous melodies and harmonies. In nature, complex structures in space and time are also not readily perceived as deriving from the same laws of harmonization, which are universal and relatively simple. Basic simplicity can form deep and wide layers of the diversity of the world. In the balance between strict periodicity and freedom of phase shift lies the essence of synchronization of complex dynamical systems. A feature of such systems is that, for all their stability, they do not repeat and are not determined for all their regularity. In this paradox lies the essence of the world around us and ourselves.

Complex dynamical systems depend on the initial conditions and current perturbations ("butterfly effect"). Lorentz equations, which started the chaos theory, showed how even a very simple nonlinearity with a limited set of variables gives rise to complex behavior. Lorenz's article was called "Predictability: Does the Flap of a Butterfly's Wings in Brazil Set Off a Tornado in Texas?" (Lorenz, 1972). It was about small perturbations, which can lead to a cardinal change in the state of the system and make it unpredictable, but obeying regularities, and, therefore, subject to forecast. Prediction is a term from rigid and deterministic views of the world; the forecast is from dynamic and probabilistic approach.

The success of the calculations for the movements of celestial bodies led to Laplace's determinism, who thought that if we calculate the state of the Universe now, we can calculate its state at any point in the future. He mixed up the prediction with the forecast. Indeed, if one puts the measurements in simple formulas of classical mechanics, one can obtain the same values or the same initial conditions. In real life, everything is not quite so. The solutions are approximate, and the variables are never repeated.

But this does not mean that knowledge and prognosis are impossible. Cognition is the forecast. And the shorter the period, the more accurate the forecast. In the long run, signal changes accumulate, and the butterfly can start a tornado. Not every Brazilian butterfly and not every flap of its wings gives rise to a tornado; otherwise, there would be no butterflies, nor us. But the sensitivity to even small changes leads to the fact that despite all the efforts of meteorology as a science and all the investments made in it by the society, which hoped for weather prediction, our forecasts remain probabilistic.

The optimism of classical mechanics was so strong that it reached the twentieth century. For example, hopes for weather prediction gave rise to hopes for climate control. In the 1950-60s, it was thought that humankind would free itself from the arbitrariness of the elements, becoming the overlord: the fields would be covered with geodetic caps, the planes would clear the sky from clouds, the scientists would understand the mechanism of starting and stopping rain, and tornadoes would no longer bother us. Of course, we influence the climate, as our activity is one of the factors, but management implies control. To control a complex dynamical system that depends on many parameters, we must control all of these parameters, each "flap of the butterfly's wings." Is it possible?

The butterfly effect seems obvious for complex systems. We all intuitively know that small shifts can lead to large earthquakes. But the fact is that the same thing works for systems that are deceiving in their simplicity. It is difficult to say whether there are non-chaotic systems in the strict sense of the word, i.e., linear and entirely predictable. Even the most periodic oscillations, such as sunrise and sunset, are only the maximum approximation to predictability and only over a certain, albeit relatively large period. The more accurate the measurements, the greater the predictability. But neither accuracy nor predictability is absolute. This is the main lesson: the absolute absence of absolutes.

What seems absolute to us, for example, the movement of the Sun, the speed of light, or something else, depends on the scale of measurement and on the time horizon within which we predict the state of the system. We can choose the parameters for assessing the state and time range, and the representation of the state space of the system will depend on this. We can describe this state space of the system and even forecast a further trajectory.

But as soon as the model includes a parameter taken as an absolute, this model signs a sentence for itself. Sooner or later, this sentence will be carried out by the very process of cognition. This is what the geometric representation of a complex dynamical system shows — a strange attractor. It can wander through states' space and even stay close to some trajectory but never repeats itself with perfect accuracy. But this does not mean that there are no patterns in it, and it is impossible to analyze it. Although the system does not repeat its states exactly, it circles around a particular attractor. This is the order of chaos.

Although a chaotic system has an attractor in the state space, the definition of the point where the system will be on the attractor in the future becomes more complicated the longer the time range and the more degrees of freedom. If we look at two such systems, then even with one type of attractor, the phase point of one will not be connected with the other system's point. But synchronization is the coupling of systems in phase space. Can chaotic systems be synchronized? In contrast to the periodic oscillators (for example, pendulums), the synchronization of stochastic oscillations seems unlikely due to their local instability as sync requires the coincidence of some parameters.

A team of researchers studied sync in chaotic systems and published their results in an article with the corresponding title "Fundamentals of synchronization in chaotic systems, concepts, and applications" (Pecora et al., 1997). First, they simulated the behavior of such systems on a computer. To the experimenters' surprise, the parameters began to converge, i.e., systems were synchronized. Then the same thing happened with real oscillatory systems. The authors wrote: "We can get an idea of what the geometry of the synchronous attractor looks like in phase space":

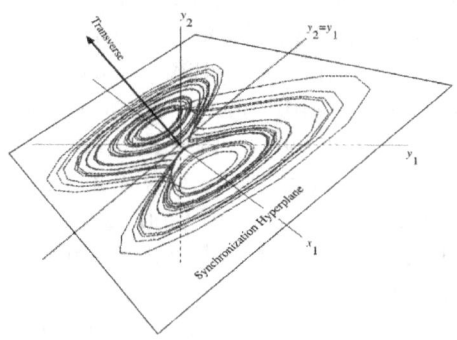

Nonlinear systems experts Nikolai Rulkov and Clifford Lewis wrote: "Studies of the generalized synchronization of chaos in the entire range of the synchronization zone must deal with non-differentiable continuous synchronization mappings. These mappings have rather complicated form and can behave differently depending upon the regime of synchronization" (Rulkov, Lewis, 2001). As an example, the authors took a synchronization regime with a frequency ratio of 2:1, which was previously observed in experiments with two chaotic electric circuits. On the phase space map, it showed the corresponding mapping in the form of a lemniscate:

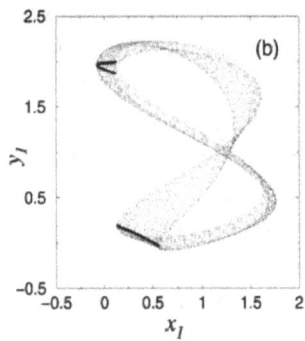

This phase portrait of an octave harmony as the base of a stable system of interacting oscillators has been with us throughout this study. We came across it when we unraveled the secrets of fundamental structures of matter in the previous volume. It will surface throughout the further volumes when we turn to biological systems and processes occurring in them, including the most mysterious one — the mind.

Here we must stress that "all roads" lead to synchronization as the interaction mechanism, be it simple pendulums, fundamental levels of matter, or complex systems including biological ones. The determining factors for the stability of any system and the corresponding coherence of the phase portrait are the frequency-phase ratios, which must be within certain limits of the synchronization region around the harmonic combinations. Harmony creates order out of chaos.

Steven Strogatz wrote: "Until just a few years ago, the study of synchrony was a splintered affair, with biologists, physicists, mathematicians, astronomers, engineers, and sociologists laboring in their separate fields, pursuing seemingly independent lines of inquiry. Yet little by little, a science of sync has begun coalescing out of insights from these and other disciplines … Those of us working in this emerging field are asking such questions as: How exactly do coupled oscillators synchronize themselves, and under what conditions? When is sync impossible and when is it inevitable? What other modes of organization are to be expected when sync breaks down? And what are the practical implications of all that we're trying to learn? … What makes these puzzles so much fun is that they lie at the edge of known mathematics … Even with the help of supercomputers, the collective behavior of gigantic systems of oscillators remains a forbidding terra incognita" (Ibid).

But who or what can stop us from exploring this uncharted territory? Even fear can only suspend curiosity, but not defeat it. If we get out of the muddle of mainstream theories and wash away the mud of their prohibitive dogmas, nothing can stop us.

One can avoid questions and follow the beaten path. One can wait for others to create something new to reap the benefits of their efforts. But there are always pioneers who boldly venture into terra incognita. They are willing to risk getting lost in the unknown. They may be misunderstood, criticized and even ignored. New ideas have always met defensive resistance from established models. This is a normal struggle between two characteristics of any model of reality: stability and dynamism. The new can become a bifurcation point and a transition to another trajectory, and it will look like a revolution. The system can also absorb changes without any perturbation, and it will look like an evolution. In the long term, what matters is the dynamic adaptability and adequacy of our model of reality, of which science is a part.

Before moving on to the next parts of our study, which will concern biophysics in general and the mechanisms of human consciousness in particular, let us summarize the main hypotheses regarding the mechanism that serves all levels of energy interactions and forms all levels of matter.

The Universe is a continuum of energy oscillations with various amplitude-frequency characteristics and phase portraits. Cross-frequency and phase coupling of these oscillations at harmonic ratios establishes stable systems by binding energy within the synchronization region. When these oscillations desynchronize, the structures disintegrate and release energy with the potential to create new harmonies. Thus, the formation or disintegration of matter is a process of synchronization and desynchronization of oscillations in an all-encompassing energy environment.

This mechanism works on all levels of matter from macrocosmic to microcosmic. The celestial bodies form dynamic ensembles of self-sustained oscillators that have coherent phase portraits with stable attractors. The harmonious frequency-phase coupling is behind the structures of galaxies, planetary or satellite systems. All of the observed phenomena of their dynamics and trajectories are the result of oscillatory interactions via waves in the all-encompassing medium. This medium can be described as an active oscillatory system that allows waves to propagate through it at great speeds over vast distances without significant energy loss. Such superconductivity features can be explained by the effective synchronization of waves with optimal harmonic ratios.

The same mechanism of frequency-phase coupling works at the microcosmic level (atomic and subatomic). The elementary structures of matter are not discrete objects (particles, material points) but continuous oscillatory processes in the all-encompassing energy environment that create wave packets with stable and coherent phase portraits that we can measure discreetly (quantize). All of the experimentally observed integer values (quantum numbers) of the fundamental interactions at this level are explained by the tendency of synchronized structures to harmony (integer ratios). All of the phenomena of attraction, repulsion, fusion, decay or stability are due to this process of synchronization and desynchronization.

At the next level of chemical elements, substances and compounds, the laws of harmony are the same. They also explain the periods in the table of chemical elements. Taking the musical analogy that is based on the same laws, we can call them the transfer by an octave of the intervals of the music of matter while preserving the harmonious combinations of frequencies. As a ratio of frequencies with a broad sync region, an octave is the fundamental periodic principle with a stable phase portrait in the form of a lemniscate ∞ (coherent and iterative trajectory). Periods have various sets of frequencies (notes of matter) that determine the difference of the resulting chemical elements (chords of matter). But all of the intervals within the octave have an inversion: no matter how the notes of matter are rearranged the harmonic frequency ratio remains the same (prima, second, third, fourth, fifth, sixth, seventh, octave) thus producing the periodic repetition of the characteristic properties of elements.

The same mechanism underlies all types of energy interactions. The only difference between various interactions is the amplitude-frequency range. Energy is transferred by waves in a continuous medium, so the manifestations of interactions depend on the parameters of the source, the medium, the receiver, and the distance. Attraction, repulsion, and the equilibrium state of interacting systems are manifestations of frequency coupling and phase relationships of energy oscillations. Since synchronization is a universal mechanism, all energy interactions can be combined within one model that describes and explains its manifestations. Thus, the *Theory of Energy Harmony* can become part of a new branch of theoretical physics, which we propose to call *Synchronology*.

REFERENCES

Arnault, A., Nicole, P. (1664). *La logique ou l'art de penser.* ch. XIX, § 3 Paris.

Awrejcewicz, J., Kudra, G. (2014). *Mathematical modelling and simulation of the bifurcational wobblestone dynamics.* Discontinuity, Nonlinearity and Complexity, 3(2), 2014, 123-132.

Barbour, J. (1999). *The End of Time: The Next Revolution in our Understanding of the Universe.* Oxford Univ. Press.

Bjerknes, V. (1906). *Fields of force.* General Books.

Blandford, R. D. (2015). *A century of general relativity: Astrophysics and cosmology.* Science 06 Mar 2015: Vol.347, Issue6226, pp.1103-1108.

Bodanis, D. (2000). *E=mc2: A Biography of the World's Most Famous Equation.* Published October 1st 2000 by Berkley Trade

Capra, F. (1975). *Tao of physics: an exploration of the parallels between modern Physics and eastern Mysticism.* Shambhala Publications, Inc.

Cavagna, Andrea, Jelic, Asja. (2014). in press http://www.sciencemag.org/news/2014/07/how-bird-flocks-are-liquid-helium.

Chladni, E. F. F. (1787). *Entdeckungen über die Theorie des Klanges.* Leipzig 1787, Reprint 1980

Corrêa, Raul. (2015) In press. *Quantum Cheshire Cat effect may be explained by standard quantum mechanics.* Lisa Zyga, Phys.org. JUNE 8, 2015

Denkmayr, T. et al. (2014). *Observation of a quantum Cheshire Cat in a matter-wave interferometer experiment.* Nat. Commun. 5:4492 doi: 10.1038/ncomms5492.

Descartes, R. (1637). *Discours de la méthode (Discourse on the Method). 1637. Meditationes de prima philosophia (Meditations on First Philosophy). 1641. Principia philosophiae (Principles of Philosophy). 1644.*

Dyson, F.J. (2007). *Why is Maxwell's Theory so hard to understand?* Conference: Antennas and Propagation, 2007. EuCAP 2007. 1 - 6. 10.1049/ic.2007.1146.

Eberlein, Claudia. (1996). *Sonoluminescence as quantum vacuum radiation.* Phys. Rev. Lett. 76, 3842.

Eberlein, Claudia. (1996). *Theory of quantum radiation observed as sonoluminescence.* Phys. Rev. A 53, 2772.

Einstein, A. (1909). *On the development of our understanding of the nature and composition of radiation.* Physikalische Zeitschrift Vol. 10. No. 22, pg. 817.

Einstein, A. (1910). *The principle of relativity and its consequences in modern physics.* Archives des sciences physiques et naturelles. 29 (1910): 5-28; 125-144

Einstein, A. (1914). *Vom Relativitdts-Prinzip.* Vossische Zeitung, 1914, 26, April, 33, 34. 395

Einstein, A. (1918). *Dialog about Objections against the Theory of Relativity.* Die Naturwissenschaften 6 (1918): 697–702.

Einstein, A. (1920a). *Ether and the Theory of Relativity.* Methuen & Co. Ltd, London, 1922.

Einstein, A. (1920b). *Fundamental ideas and methods of the Theory of relativity, presented in their development.*

Einstein, A. (1926). *letter to N. Bohr, 12 December 1926,* in Ronald William Clark. *Einstein The life and times.* Published January 1st 2001 by Avon (first published January 1st 1971).

Einstein, A. (1954) *Ideas and Opinions.* Crown Trade Paperbacks

Einstein, A., Minkowski, H. (1920). *The Principle of Relativity.* University of Calcutta.

Einstein, A., Podolsky, B., Rosen, N. (1935). *Can quantum-mechanical description of physical reality be considered complete?* Phys. rev./ G.D. Sprouse American Physical Society Vol. 47.

Einstein, A., Infeld, L. (1938). *The Evolution of Physics.* Edited by C.P. Snow, Cambridge University Press.

Epstein, Irving R. (2006). *Predicting complex biology with simple chemistry.* PNAS 2006 103 (43).

Faraday, Michael. (1844). *A Speculation touching Electric Conduction and the Nature of Matter.* The London, Edinburgh, and Dublin Philosophical Magazine and Journal of Science, 3, 1844, pp. 136-144.

Feynman, Richard P.; Leighton, Robert B.; Sands, Matthew (1964-1966) T*he Feynman Lectures on Physics.* The Definitive and Extended Edition (2nd ed.). Addison Wesley. 2005

Feynman, Richard (1965). *The Character of Physical Law.* MIT Press.

Feynman, R. (1985). *QED: The Strange Theory of Light and Matter.* Princeton.

Feynman, R. (1988). *What Do You Care What Other People Think.* New York: W. W. Norton.

Flannigan, DJ, Suslick, KS. (2005). *Plasma formation and temperature measurement during single-bubble cavitation.* Nature. 434 (7029): 52–5. March 2005.

Forbes, George. (1881). *Hydrodynamic Analogies to Electricity and Magnetism.* Nature, 15 August 1881

Gribbin, J. (2000). Introduction. Q is for QUANTUM: An Encyclopedia of Particle Physics. C. 7.

Hawking, S. (1988). *A brief history of Time from the Bing Bang to Black Holes.* Bantam Dell Publishing Group.

Heisenberg, W. (1958). *Physics and philosophy.* New York: Harper & Row Русский перевод: В. Гейзенберг. Физика и философия. Часть и целое. Наука, 1990.

Hotson, D. (2002). *Dirac's Equation and the Sea of Negative Energy (PART 1,2).* Infinite Energy Issue 43, 44.

Hotson, D. (2009). *Dirac's Equation and the Sea of Negative Energy (PART 3).* Infinite Energy Issue 86.

Jenny, Hans (1967). *Kymatik – Wellen und Schwingungen mit ihrer Struktur und Dynamik.* Cymatics – The Structure and Dynamics of Waves and Vibrations.

Kakalios, James. (2005). *Resource Letter GP-1: Granular physics or nonlinear dynamics in a sandbox.* American Journal of Physics 73, 8 (2005); doi: 10.1119/1.1810154

Krisch, A.D. (1979). *The spin of the proton.* Scientific American, May 1979, p. 68-80.

Krisch, A.D. (2005). *Violent Collisions of Spinning Protons.* XIth International Conference on Elastic and Diffractive Scattering, Chateau de Blois, France, May 2005 (arXiv: hep-ex/0511040)

Kuramoto, Y. (1975). *International Symposium on Mathematical Problems in Theoretical Physics.* Springer Lecture Notes Physics, Vol. 39, edited by H. Araki (New York: Springer), p. 420.

Laughlin, R. B. (2005). *A Different Universe: Reinventing Physics from the Bottom Down.* NY, NY:: Basic Books.

Leacock, Stephen. (1956). *Common Sense and the Universe.* James R. Newman *The World of Mathematics Volume 1* George Allen & Unwin 1956.

Lorenz, E. N. (1972). *Predictability: Does the Flap of a Butterfly's Wings in Brazil Set Off a Tornado in Texas?* . American Association for the advancement of science.

Maxwell, J. (1855). *On Faraday's Lines of Force.*

Maxwell, J. (1861). *On physical lines of force.* Philosophical Magazine. 90: 11–23.

Maxwell, J. (1873). *A Treatise on Electricity and Magnetism.* Oxford University Press, London.

Maxwell, J. (1878). *The Encyclopaedia Britannica. A dictionary of arts, sciences, and general literature* Ninth Edition Volume VIII.

Melo, Francisco, Umbanhowar, Paul, Swinney, Harry. (1994). *Transition to Parametric Wave Patterns in a Vertically Oscillated Granular Layer.* Physical review letters. 72. 172-175. 10.1103/PhysRevLett.72.172.

Melo, Francisco, Umbanhowar, Paul, Swinney, Harry. *(1995). Hexagons, kinks, and disorder in oscillated granular layers.* PHYSICAL REVIEW LETTERS 1995 Nov 20;75(21):3838-3841.

Mindell, A. (2000). *Quantum Mind: The Edge Between Physics and Psychology.* Portland, OR: Lao Tse Press.

Moran, M.J., Haigh, R.E. et al. (1995). *Direct observations of single sonoluminescence pulses.* Nuclear Instruments and Methods in Physics Research Section B: Beam Interactions with Materials and Atoms, Volume 96, Number 3, May 1995, pp. 651-656(6)

Newton, I. (1687). *The Principia: Mathematical Principles of Natural Philosophy.* University of California Press, 1999.

Newton, I. (1704). *Opticks: or, a treatise of the reflexions, refractions, inflexions and colours of light. Also two treatises of the species and magnitude of curvilinear figures.* Palo Alto, Calif.: Octavo. 1998

Newton, I. (1756). *Four letters from Sir Isaac Newton to Doctor Bentley, containing some arguments in proof of a deity.* London: Printed for R. and J. Dodsley, 1756.

Newton, I. (1992). *Newton's Philosophical Writings.* Cambridge University Press.

Oppenheimer, R. (1949). Quoted in "J. Robert Oppenheimer" by L. Barnett, in Life, Vol. 7, No. 9, International Edition (24 October 1949), p. 58.

Pecora, Louis M., Carroll, Thomas L., Johnson, Gregg A., Douglas, J. (1997). *Fundamentals of synchronization in chaotic systems, concepts, and applications.* Chaos 7 (4).

Peterson, Ivars. (1990). *Proton spin plays key role in smash hits – collisions between subatomic particles.* Science News, Nov 3, 1990

Petrov, V., Gaspar, V., Masere, J., Showalter, K. (1993). *Controlling chaos in the Belousov-Zhabotinsky reaction.* Nature 361:240-3.

Pikovsky, A., Rosenblum, M. (2003). *Synchronization: from pendulum clocks to chaotic lasers and chemical oscillators.* Contemporary Physics, September – October 2003, volume 44, number 5, pages 401 – 416.

Pikovsky, A., Rosenblum, M., Kurths, J. (2001). *Synchronization: A Universal Concept in Nonlinear Sciences (Cambridge Nonlinear Science Series).* Cambridge University Press.

Poincaré, H. (1900). *La théorie de Lorentz et le principe de réaction. Archives néerlandaises des sciences exactes et naturelles. Vol. 5. — P. 252—278.*

Popular Science Monthly (1882). Volume 21. June 1882. *Hydrodynamics and Electricity.*

Rayleigh, J. (1945). *The Theory of Sound.* New York: Dover Publ.

Rosenthal-Schneider, Ilse (1980). *Reality and Scientific Truth.* Detroit: Wayne State University Press, 1980. p 74.

Roy, R., Thornburg, K. S. . (1994). *Experimental synchronization of chaotic lasers.* Phys. Rev. Lett., 72:2009–2012

Rulkov, N. F., Sushchik, M., Tsimring, L.S., Abarbanel H. D. I. . (1995). *Generalized synchronization of chaos in directionally coupled chaotic systems.* Phys. Rev. E, 51(2):980–994.

Rulkov, Nikolai F., Lewis, Clifford Tureman. (2001). *Subharmonic destruction of generalized chaos synchronization.* Physical Review Volume 63 The American Physical Society.

Sacks, O. (1998). *Understanding. A Leg to Stand On.* Simon and Schuster.

Shattuck, M.D., Bizon, C., Umbanhowar, P.B., Swift, J.B., Swinney, H.L. (1998) *Pattern Formation in Vertically Vibrated Granular Layers: Experiment and Simulation.* In: Herrmann H.J., Hovi JP., Luding S. (eds) Physics of Dry Granular Media. NATO ASI Series (Series E: Applied Sciences), vol 350. Springer, Dordrecht

Silva, Cibelle. (2006). *The Role of Models and Analogies in the Electromagnetic Theory: A Historical Case Study.* Science & Education. 16. 10.1007/s11191-006-9008-z.

Smolin, L. (2006). *The trouble with physics. The Rise of String Theory; the Fall of a Science, and What Comes Next.* Houghton Mifflin.

Strassler, M. (n.d.). *Virtual Particles: What Are They?* Available online at: https://profmattstrassler.com/articles-and-posts/particle-physics-basics/virtual-particles-what-are-they/.

Strogatz, S. (2003). *Sync: The Emerging Science of Spontaneous Order.* Hyperion.

Sussman, G. J., Wisdom, J. (1988). *Numerical evidence that the motion of Pluto is chaotic.* Science, 241, 433–437.

Taleyarkhan, R. P., West C.D., Lohey R.T., Nigmatulin R.I., Block R.C. (2002*). Evidence for nuclear emissions during acoustic cavitation.* Science 295, 1868–1873.

Taylor, J. (2001). *Hidden Unity in Nature's Laws* . Cambridge University Press, New York.

Taylor, R. L. (2011). *Attractors: Nonstrange to Chaotic.* Conference proceedings.

Thomson, W. (1891). *Popular Lectures and Addresses. Vol. 1* London: MacMillan. p. 80. *Retrieved 25 June 2012.*

Umbanhowar, Paul B., Melo, Francisco, Swinney, Harry L. (1996). *Localized excitations in a vertically vibrated granular layer.* Nature 382, 793 – 796 (29 August 1996)

Velikhov, A.N. (2003) *Meeting of the RAS Presidium.* Bulletin of the Russian Academy of Sciences. Volume 73. No. 4. 2003.

Whittaker, E. T. (1910). *A history of the theories of aether and electricity from the age of Descartes to the close of the nineteenth century.* Longmanns, Green and Co.

Weinberg, S. (2013). *Lectures in Quantum Mechanics.* Cambridge University Press.

Weyl, H. (1927). *Philosophie der Mathematik und Naturwissenschaft.*

Winfree, A. (1980). *The Geometry of Biological Time.* Springer, New York (Reprinted as Springer Study Edition, 1990, Springer, Berlin).

Books in the Series
Symphony of Matter and Mind

Part one

Music of Matter
Mechanism of Material Structures Formation

Part two

Theory of Energy Harmony
Mechanism of Fundamental Interactions

Part three

Music of Life
Physics and Technology of Living Matter

Part four

Algorithm of the Mind
Teleological Transduction Theory

Part five

Technologies of the Mind
The Brain as a High-Tech Device

Part six

Harmonies of the Mind
Physics and Physiology of Self

Part seven

Inner Universe
The Mind as Reality Modeling Process

Part Eight

Dissonances of the Mind
The Physics of Mental Disorders

About the Author

Stanislav Tregub

Independent researcher.

Research areas: physics, biophysics, neuroscience, psychology, psychiatry.

www.ingramcontent.com/pod-product-compliance
Lightning Source LLC
Chambersburg PA
CBHW082105220526
45472CB00009B/2056